Introduction

I am pleased to present another title in the "Cattle" series.

The work is in the Public Domain and is re-printed here in accordance with Federal Laws.

As with all reprinted books of this age that are intended to perfectly reproduce the original edition, considerable pains and effort had to be undertaken to correct fading and sometimes outright damage to existing proofs of this title. At times, this task is quite monumental, requiring an almost total "rebuilding" of some pages from digital proofs of multiple copies. Despite this, imperfections still sometimes exist in the final proof and may detract from the visual appearance of the text.

I hope you enjoy reading this book as much as I enjoyed making it available to readers again.

Jackson Chambers

DIGEST

1. TESTS OF DAIRY COWS, 1908-1909.

The Number of Official Tests of Dairy Cows during the year ending October 1, 1909, aggregated 1479, an increase of 11 per cent over the previous year. The various breeds were represented as follows: Holstein cows 364, Guernseys 143, Jerseys 28, Red Polled 7, grade 1, a total of 543. Tests of Holstein cows were conducted as follows: 20 2-day tests, 532 7-day tests, and 10 30-day tests. These tests were conducted on 364 cows owned by 42 different breeders. Wisconsin Holstein breeders secured a large number of the Holstein-Friesian Association prizes including one first, three each of second and third and 16 other prizes, or 11.2 per cent of the total number of prizes offered by the Association this year. Tests of Guernsey cows were all semi-official and made in cooperation with the American Guernsey Cattle Club. A total of 919 monthly tests were made on 143 cows and a large number of them qualified for the advanced registry of the A. G. C. C. Tests of Jersey cows included 24 semi-official tests and three 7-day tests, the former under the American Jersey Cattle Club rules for admission to their Register of Merit. Tests of Red Polled and Grade cows included 10 2-day and two 3-day tests.

II. A DECADE OF COW TESTING.

Official Tests of Dairy Cows have been Conducted in Wisconsin since the early nineties. During the last decade these tests have been conducted in 27 counties for 109 breeders on a total of 2,764 cows. The benefits of the tests have not been confined to this number of breeders, however, since the system of official and semi-official testing has been largely responsible for the improved breeding of dairy stock, which has exerted an influence upon the entire dairy industry. The tests have been particularly valuable in enabling breeders to furnish accurate figures of the production of their animals and in educating the public to the economic importance of good blood with proper feeding, and care for dairy animals.

The Growth in Number of Tests has been steady for the last nine years, the most rapid increase occurring during the last two years. A marked improvement in the production of cows of the various classes has been noted. During the first five years of the decade, the average production of aged Holstein cows on 7-day tests was 397.5 pounds milk and 13.9 pounds butter fat, while in the last five years of the decade the average production of this class was 432.9 pounds milk and 15.3 pounds butter fat. This improvement is attributed to better breeding and good feed and care given to the cows.

Several Cows of Special Merit have been tested during this 10-year period, notably the Guernsey cow Yeksa Sunbeam, 15439, the Guernsey heifer Yeksarose, 16610, the Jersey cow Double Time, 157531, and the Holstein cow Colantha 4th's Johanna, 48577. All of these cows stand among the foremost in their respective classes for the various breeds.

A Decade of Official Tests of Dairy Cows, 1899-1909.

F. W. WOLL AND R. T. HARRIS.

The official testing of dairy cows for their production of milk and butter fat during stated periods has gradually grown in importance in Wisconsin during the past dozen years or more, both to breeders of pure-bred cattle in the state for whom the tests are conducted, and to this Experiment Station under whose direction the work is done. This work assumed large proportions after the establishment of the Advanced Registers of the three main dairy breeds in this state, viz., by the Holstein-Friesian Association of America in 1894 and the American Guernsey and Jersey Cattle Clubs in 1901 and 1903, respectively, when the records made in these tests became of direct money value to the owners of the cows. The value of an animal and her near relatives is increased considerably by her being admitted to the Advanced Register of her breed. At a very low estimate this increase is at least $25 per head. Official testing of dairy cows, the main condition for admittance to the Advanced Registers, thus becomes a business proposition for the owner. He is primarily benefited by the tests and under the rules and schedule of rates now in force, he pays nearly the entire expense of the tests to this Station. The justification for our continuing to perform this service to the individual breeders and breed associations lies in the fact that the tests are also of general value to the dairy interests of the state. Dairy farmers are dependent, as a rule, upon the breeders of pure-bred cattle to supply the sires to head their herds, and at times buy females as well. A systematic and accurate determination of the productive capacity of a pure-bred dairy cow is for this reason of importance both to the breeder in the conduct of his dairy herd and to his customers among the dairymen of the state, hence, in general, to the dairy industry of the state.

One of the conditions of conducting the tests is that the records shall be public property and may be used for the study of relations bearing on the production of milk and butter fat. During the past 10 years accounts of the results of the tests have been published in bulletin form or in the annual reports of the Station,[1] and the present testing year closes a period of 10 years during which the tests have been conducted under the direction of the writer (W.). It has been considered well, therefore, in addition to giving the usual report on the work done in this line during the past year, to review briefly the results obtained during the 10-year period stated, 1899 to 1909, inclusive, and to discuss these results in their practical bearings to the extent that is deemed advisable. The main results of the testing work done during the past year are presented in the following pages in the same manner as has been the custom in earlier years, and summaries for the past 10-year period with discussions are given in the second part of this bulletin. The history of dairy tests in this state was given by the writer in Bulletin 107 of this Station and will not be repeated here, as that bulletin is still available for distribution. In the same place also will be found a general discussion of the method of conducting official tests and the rules governing the same which have been in force during the last 10 years.

I. TESTS OF DAIRY COWS, 1908–1909.

The practice followed during the last two years of publishing only records of production that would entitle cows to special consideration has been continued in this bulletin. During the past year, 1,479 tests were conducted in all, of 543 different cows, against 1,327 tests of 418 cows during the preceding year, an increase of 11 per cent in the number of tests conducted and of 30 per cent in the total number of cows tested. Table I shows the number of tests conducted during the year ending

---

[1] Rept. 17 (1900) pp. 62 to 75; Rept. 18 (1901) pp. 73 to 84; Rept. 19 (1902) pp. 107 to 127; Bul. 107 (Dec. 1903) 43 pp. Rept. 21 (1904) pp. 112 to 142; Bul. 131 (Dec. 1905), 46 pp.; Rept. 22 (1905) pp. 125 to 128; Bul. 144 (Jan. 1907), 65 pp.; Rept. 24 (1907) pp. 76 to 88; Bul. 160 (Mar. 1908), 39 pp.; Bul. 172 (Mar. 1909), 33 pp.

October 1, 1909, with the number of cows tested during this period and in the aggregate since this work was first taken up by this Station in a systematic manner.

TABLE I.—SUMMARY OF TESTS CONDUCTED AND OF COWS TESTED, 1908-9, AND 1893-1909.

	1908-1909.			1893-1909.
	No. of tests.	No. of cows.	Total per breed.	No. of cows.
Holstein, 2-day tests..........	20	5	364	2283
7-day tests..........	352	349		
30-day tests..........	10	10		
Guernsey, 1-day tests..........	919	143	143	702
Jersey, 2-day tests..........	161	25	28	171
7-day tests..........	3	3		
Ayrshire, 7-day tests..........	....	....	....	3
Shorthorn, 7-day tests..........	....	....	....	12
Brown Swiss, 7-day tests..........	....	....	....	9
Red Polled, 2-day tests..........	10	5	7	41
3-day tests..........	2	2		
Grades, 1-day tests..........	2	1	1	43
Totals..........	1479	543	543	3264

Of the number of tests conducted last year, 365 were official 7-day or 30-day tests, while 1,114 were 1-day or 2-day tests conducted for semi-official yearly records. The corresponding figures for the preceding year were 263 and 1,064 tests, respectively, showing an increase of about 39 and 5 per cent in the number of tests of the two kinds conducted during 1908 to 1909, as compared with the preceding year. The number of cows of the various breeds tested during the year were as follows: Holstein, 364; Guernseys, 143; Jerseys, 28; Red Polled, 7; and 1 grade, making a total of 543 cows. During the past 17 years we have tested in all over 3,260 cows, of which number 2,283 were Holsteins, 702 Guernseys, 171 Jerseys, 43 grades, 41 Red Polled, 12 Shorthorns, 9 Brown Swiss, and 3 Ayrshires.

The results given in this bulletin have been grouped according to the breeds of the cows, in the order of the number of cows tested of each breed, viz., Holstein, Guernsey, Jersey, Red Polled, and grades. Photographs of some of the best cows tested during the year are reproduced in connection with the tables, from prints furnished by the owners of the cows. The illustrations, as a rule, give a good idea of the conformation and general appearance of these cows. It proved impossible in many cases to obtain photographs of cows whose production during the past year would entitle them to be represented in this bulletin.

TESTS OF HOLSTEIN COWS.

Some 382 different tests of Holstein cows were conducted during the year, viz., 20 2-day tests, 352 7-day tests, and 10 30-day tests. Many of the 7-day tests were continued for several days beyond that limit, and a few were continued for 14 days. The results of these miscellaneous tests, although generally reported to the Superintendent of Advanced Registry of the Holstein-Friesian Association, are not given in this bulletin.

SEVEN-DAY OFFICIAL TESTS. As in earlier years the large majority of official tests of Holstein cows conducted by this Station were of seven days' duration and were made at the beginning or the early part of the lactation of the cows. Some 346 7-day tests at the beginning of the lactation period, of 343 different cows, and 6 tests at about 8 months from the time of calving were conducted during the year. The 364 cows tested were owned by 42 different breeders, all but two located in this state. The names and addresses of the owners, with the number of cows tested for each, are given in Table II.

TABLE II.—OWNERS OF HOLSTEIN COWS TESTED DURING THE YEAR ENDING OCTOBER 1, 1909.

No.	Name and address.	No. of cows.	No.	Name and address.	No. of cows.
1	F. W. Allis, Madison	45	23	August Luedke, Juneau	4
2	E. E. Ayer, Walworth	2	24	J. L. Mason, Elgin, Ill	1
3	Albert Babler, Jr., Monticello	18	25	C. H. & W. B. Nevens, Winnebago, Ill	6
4	F. J. Bristol & Sons Co., Oakfield	21	26	A. G. Palmer, Lake Geneva	1
5	M. W. Demerit, Lake Mills	5	27	M. F. Peck & Sons, Marshall	5
6	H. D. Dunbar, Elkhorn	1	28	E. C. Petrie, Elkhorn	7
7	John Erickson, Waupaca	8	29	F. E. Randall, Watertown	12
8	Wm. Everson & Sons, Lake Mills	5	30	S. M. Randall, Waupun	5
9	L. W. Gay, Madison	6	31	C. B. Reddelien, North Lake	1
10	Edwin D. Gibbs, Fox Lake	2	32	R. J. Schaefer, Appleton	8
11	H. B. Giddings, Sheboygan Falls	4	33	C. A. Schroeder & Son, West Bend	8
12	W. J. Gillett, Rosendale	7	34	Ed. M. Schultz, Hartford	9
13	John Hetts, Fort Atkinson	5	35	Geo. J. Schuster, Mukwonago	7
14	Imig Bros., Neillsville	7	36	C. H. Stevens, Jefferson	15
15	Nim Johnson, Elkhorn	3	37	W. F. Tidyman, Oakfield	8
16	S. B. Jones & Son, Watertown	21	38	University of Wisconsin, Madison	3
17	W. H. Jones, Watertown	2	39	Chas. Van der Schaaff, Sparta	6
18	A. P. Kaye, Walworth	2	40	A. L. Williams, Fond du Lac	8
19	L. A. Kimball, Lake Geneva	2	41	Tompkins Wright, Waupun	10
20	August Knospe, Juneau	13	42	Geo. Young, Reedsburg	2
21	Fred Kohlwey, Grafton	2			
22	L. G. Legler, Juda	36			

The tests of the three cows owned by the University were conducted under our supervision, and the composite samples and records, on completion of the tests, were submitted for verification to Prof. W. J. Fraser, Chief of the Department of Dairy Husbandry of the University of Illinois. The composite samples of milk received on the various tests during the year were tested in the dairy laboratory of this Station by either A. C. Oosterhuis or E. H. Hayes, the large majority of the samples having been tested by the former assistant.

In conducting these tests the Station employed 17 young men, graduates of one of the courses in the College of Agriculture of this University. The Supervisors of Dairy Tests are selected from the eligible list furnished by the State Civil Service Commission, which is made up according to the standings of the various candidates in examinations held by the Commission. The names of the supervisors during the past year, with the number of Holstein cows on official 7-day tests conducted by each supervisor are given in Table III.

TABLE III—SUPERVISORS OF OFFICIAL 7-DAY TESTS OF HOLSTEIN COWS.

No.	Name and address.	No. of cows	No.	Name and address.	No. of cows
1	George A. Allen, Winneconne...	4	10	W. H. Markey, Sullivan.........	34
2	E. L. Dreger, Madison...........	22	11	H. M. Peck, Marshal............	23
3	C. N. Dunham, La Crosse........	1	12	Otto A. Reinking, Madison......	3
4	Geo. A. Durnford, Rockbridge...	22	13	Wm. Schwichtenberg, Oshkosh..	35
5	Wm. P. Ensch, Mauston..........	5	14	Paul Skeflo, Madison...........	63
6	W. H. Gardner, Solon Mills, Ill..	48	15	Chas. L. Turner, Elkhorn.......	40
7	Ruthven E. Harris, Warrens.....	8	16	F. R. Weymouth, Plainfield.....	6
8	F. R. Johnston, Appleton........	30	17	Jas. R. Hatch, Waupaca.........	8
9	Robt. Lachmund, Sauk City.....	14			

The results of the more important tests conducted during the year are given in the following pages. No records have been included in this year's bulletin that did not go above the following standards of production of butter fat during seven consecutive days for cows of the various classes, viz.:

Class I (cows over five years old), 18 pounds butter fat.
Class II (four to five years old), 16 pounds butter fat.
Class III (three to four years old), 14 pounds butter fat.
Class IV (two to three years old), 12 pounds butter fat.
Class V (under two years old), 10 pounds butter fat.

TABLE IV.—OFFICIAL TESTS OF HOLSTEIN COWS, 1908-1909.

Cow No.	Name of Cow	Herd book No.	Test began.	Age. Y. M. D.	Days in milk.	Yield in seven days. Milk. Lbs.	Yield in seven days. Fat. Lbs.	Per cent fat. Average.	Per cent fat. Range.	Fat per day. Range. Lbs.	Years previously tested.
	Class I.—Cows five years old and over.										
1	Lady Longfield 4th	44125	Apr. 3	12— 1—20	16	541.9	20.821	3.84	2.93—5.0	2.754—3.197	99, 00, 01,04,05, 08
2	Johanna De Pauline 2d	48576	Dec. 19	10— 0— 7	22	511.3	20.847	4.08	3.3 —5.05	2.865—3.038	01, 02.
3	Princess of Grafton	51220	Dec. 31	9— 0—21	33	520.4	19.577	3.76	3.23—5.5	2.534—3.670	
4	Johanna De Colantha	52253	Mar. 26	8—11—12	17	581.7	20.044	3.79	2.5 —5.15	2.942—3.307	02, 05, 08.
5	Kate Spray 3d	56404	Mar. 29	8— 6—21	23	520.6	18.207	3.50	3.13—4.1	2.477—2.732	
6	Iolena Margaret	56423	Mar. 30	8— 4—27	12	472.8	18.834	3.98	3.13—4.65	2.529—2.825	08.
7	Netherland Johanna Rue 2d	58125	Jan. 7	7—11—11	23	546.6	18.964	3.65	3.0 —4.65	2.611—3.121	07, 08.
8	Johanna Rue de Kol	58549	Jan. 19	7— 8—16	4	407.1	19.144	4.70	3.68—6.7	2.398—3.141	03, 04, 05.
9	Lettie B	59021	Apr. 8	7—11—26	14	465.7	19.670	4.22	3.0 —6.1	2.510—3.526	05, 07.
10	Snowball Pink	59077	Apr. 15	8— 1— 3	8	486.1	18.671	3.84	3.15—4.25	2.540—2.959	03, 04, 05, 07.
11	Johanna De Pauline 4th	60902	Mar. 23	7— 1—28	18	582.9	20.375	3.50	2.7 —4.15	2.862—2.926	03, 04, 05, 06, 07.
12	Lucyra Colanthus Pietertje	61369	Jan. 16	7— 2—15	23	471.5	18.747	3.98	3.25—4.95	2.651—2.747	06.
13	Jessie Fobes 7th	62993	Mar. 28	7— 1— 1	7	552.9	20.501	3.71	3.2 —4.53	2.806—2.987	04, 07.
14	Jessie Fobes 6th Homestead	64246	Jun. 6	7— 0— 4	13	593.1	20.823	3.51	2.1 —4.0	2.619—3.211	04,06,07,08(22,596)
15	Jewel Duchess	64474	M'r. 1	6— 4—11	12	514.6	22.109	4.30	3.55—5.3	2.079—3.455	04, 07.
16	Little Fay	65549	Feb. 1	7— 9—19	25	47.0	18.451	3.71	3.2 —4.4	2.501—3.001	
17	Abbie Douglas De Kol 2d	65090	Dec. 22	6— 0—11	19	555.7	18.357	3.43	2.5 —4.8	2.388—2.720	05, 06.
18	Oak De Kol 2d	66793	May 3	6— 5— 1	5	547.4	23.040	4.32	3.4 —5.03	3.105—3.490	04, 06, 07.
18	Oak De Kol 2d	66793	Jun. 8	6— 5— 1	42	642.9	20.691	3.22	2.05—4.3	2.447—3.537	
19	Johanna Mercedes	66888	Jan. 21	6— 2— 2	19	507.4	18.525	3.65	3.2 —4.33	2.539—2.797	04, 07.
20	Gewina 2d's Lilly 3d	68194	Apr. 20	5— 4—20	21	540.0	18.378	3.40	2.5 —5.7	2.337—3.242	07, 08.
21	Ollie Watson Prima Donna	71767	Feb. 25	5— 3— 7	14	508.4	21.780	4.28	3.8 —4.73	2.985—3.423	07.
22	Wisconsin Bess P. ebe	71899	Dec. 14	5— 3—11	31	441.6	18.396	4.17	2.05—5.7	2.291—2.938	05, 07.
	Average, 23 tests (22 different cows)			7— 6— 0	19	**521.3**	**19.937**	**3.82**	2.05—6.7	2.291—3.670	
	Average, 112 tests (110 different cows)			7— 7— 2	23	441.8	15.590	3.53	2.0 —8.4	1.244—3.670	
	Class II.—Cows four years old and under five.										
23	Jessie Gem Inka	73592	May 6	4—11—29	39	516.0	19.674	3.81	3.0 —4.8	2.478—3.441	08.
24	Pietertje Mechthilde Beauty	74914	Mar. 7	4—10— 4	11	511.4	18.202	3.56	3.05—4.1	2.437—2.844	08.
25	Parthenea Julia Mechthilde	74918	Apr. 2	4—11— 6	7	517.9	16.800	3.24	2.6 —5.0	2.180—2.659	06.
26	Plebe Lou field Night	75249	Nov. 26	4— 5—22	10	534.3	22.118	4.14	3.0 —5.0	2.065—3.527	06, 07.

	No.	Name		Date	Age y-m-d	Days	Milk lbs.	Fat lbs.	Per cent fat	Fat range	Milk range	Year
	27	Phebe Es ato Dora	76486	Feb. 13	4— 0—11	8	441.4	16.645	3.77	3.0 —4.78	2.174—2.734	
	28	Lucyra Mercedes De Kol	77832	Dec. 11	4— 2— 0	11	481.9	16.882	3.50	2.95—3.9	2.306—2.507	06, 07.
	29	Johanna De Colantha 3d	77938	Mar. 29	4— 1—11	13	546.8	16.945	3.10	2.5 —3.65	2.376—2.470	07.
	30	Lady Oak Homestead Ormsby	78670	Jan. 1	4— 3— 5	7	391.6	16.383	4.15	2.78—4.6	2.136—2.467	
	31	Wild Rose Johanna De Kol	78898	Nov. 20	4— 0—22	12	403.8	16.393	4.06	3.48—5.05	2.126—2.648	07.
	32	Wild Rose Piebe Homestead	79500	Mar. 24	4— 5—15	11	411.1	16.039	3.90	3.13—4.58	2.253—2.344	
	33	Barbara Tirania Mechthilde	82263	Apr. 6	4— 0—20	13	481.9	16.668	3.46	2.9 —4.25	2.240—2.454	
	34	Jessie Pobes Myrtle Piebe 3d	83264	May 3	4— 0—13	30	510.6	16 6.0	3.27	2.8 —3.8	2.302—2.454	07, 08.
	35	Hendrika Girl De Kol	90882	Apr. 8	4— 2—25	20	460.7	16.804	3.65	2.9 —4.5	1.960—2.787	
		Average, 13 tests (13 different cows)			4— 4— 9	**15**	**477.9**	**17.403**	**3.64**	**2.5 —5.05**	**1.930—3.527**	
		Average, 44 tests (43 different cows)			4— 5— 6	20	410.9	14.466	3.52	2.2 —5.95	1.313—3.527	
		Class III.— Cows three years old and under four.										
	36	Aaltje Salo Ruth	76484	Jan. 28	3—11—23	13	420.7	14.885	3.54	2.9 —4.2	1.990—2.190	
	37	Pinketje Pia 3d	77930	Dec. 21	3—11—26	16	470.6	15.937	3.39	2.98—3.85	2.210—2.350	07.
	38	Aaggie Carlota Paulin	79538	Nov. 28	3—10— 2	8	409.4	14.265	3.48	2.95—4.5	1.835—2.141	06, 07.
	39	Abbie Douglas Mutual De Kol	79721	Nov. 20	3—11— 6	25	463.5	15.216	3.28	2.4 —4.2	1.976—2.308	
	40	Johanna Rue Burke	81401	Dec. 20	3— 5— 5	38	379.9	14.063	3.70	2.75—4.48	1.852—2.167	
	41	Madrigal Josephine Gerben	82908	Apr. 11	3— 3— 2	6	4 3.2	16.057	3.47	2.8 —6.7	2.047—2.784	
	42	Jessie De Kol Piebe	84316	Apr. 9	3— 4—15	7	436.3	14.912	3.27	2.5 —3.75	2.053—2.261	08.
	43	Maid De Kol Johanna	84459	Jan. 2	3— 3—23	15	441.7	14.268	3.21	2.7 —3.6	1.989—2.068	
	44	Wase Netherland Burke	86485	Nov. 27	3— 0—27	21	460.0	15.814	3.44	2.63—4.45	2.029—2.641	
	45	Uneeda Dolly Korndyke	87484	Dec. 22	3— 2—13	7	434.1	14.088	3.24	2.1 —4.4	1.902—2.138	
	46	Bessie Ward De K 1	86553	Nov. 21	3— 1—22	16	479.1	16.967	3.54	2.93—4.13	2.243—2.649	
	47	Pauline De Colantha	86678	Feb. 1	3— 1—16	19	403.0	15.225	3.78	3.4 —4.5	2.000—2.297	
	48	Johanna Clothilde 3d's Pet	87417	Jan. 14	3— 2—24	17	374.7	14.435	3.85	2.4 —6.45	1.793—2.098	07, 08.
	49	Star Watson Mooie	87770	Dec. 17	3— 4—21	19	434.9	14.358	3.30	2.38—4.0	2.003—2.076	
	50	De Kol Douglas 5th	88082	May 13	3— 5—29	9	477.9	17.200	3.60	2.85—4.3	2.346—2.584	08.
	51	Lady Longfield 4th's Homestead De Kol	89109	Apr. 26	3— 0— 6	19	484.7	16.259	3.35	2.85—4.58	2.106—2.372	08.
	52	Thoro De Brie	90432	Feb. 20	3— 9—22	29	502.3	15.963	3.18	2.4 —4.15	1.930—2.898	
	53	Madison Jennie	100698	Feb. 28	3—11—10	11	399.2	14.602	3.66	3.1 —4.2	1.983—2.165	08.
		Average, 18 tests (18 different cows)			3— 5—15	**16**	**442.1**	**15.251**	**3.44**	**2.1 —6.7**	**1.793—2.898**	
		Average, 61 tests (55 different cows)			3— 4— 16	25	377.9	12.849	3.40	1.9 —6.7	1.174—2.898	
		Class IV. Cows two years old and under three.										
	54	Inka Korndyke Burke	86484	Nov. 19	2—11—29	29	441.8	13.768	3.12	2.55—3.7	1.701—2.1 0	07.
	55	Uneeda Colantha Korndyke	86449	Nov. 20	2—10—23	25	398.4	12.017	3.02	2.5 —3.55	1.604—1.813	
	56	Uneeda Douglass Korndyke	86493	Dec. 1	2—10—23	5	388.0	14.803	3.82	3.1 —4.7	2.019—2.299	08.
	57	Aaggie Netherland Korndyke	86494	Dec. 2	2—10— 8	6	446.8	18.826	4.21	3.6 —4.6	2.546—2.779	08.
	58	Woodland Lady Burke	86496	Dec. 1	2—10— 8	27	333.6	13.311	3.99	3.0 —5.2	1.774—2.073	
	59	Inka Korndyke De Kol Burke	6497	Nov. 2	2— 9—22	18	408.0	14.346	3.52	2.3 —4.6	1.838—2.349	08.
	60	Mercedes Korndyke De Kol Burke	86495	Nov. 20	2— 9— 8	7	369.2	14.072	3.81	2.5 —5.05	1 838—2.171	

TABLE IV.—OFFICIAL TESTS OF HOLSTEIN COWS, 1908-1909—Continued.

Cow No.	NAME OF COW.	Herd book No.	Test began.	Age. Y. M. D.	Day's in milk.	YIELD IN SEVEN DAYS. Milk. Lbs.	YIELD IN SEVEN DAYS. Fat. Lbs.	PERCENT FAT. Average.	PERCENT FAT. Range.	FAT PER DAY. Range. Lbs.	Years previously tested.
	Class IV.—Cows two years old and under three										
61	Leland Sunbeam	88664	Jan. 1	2— 9—27	27	360.4	12.194	3.38	2.4 –4.43	1.664–1.867	
62	Lady Oak 2d Homestead De Kol	89107	Apr. 22	2—11—17	30	506.4	17.883	3.53	2.5 –5.55	2.169–2.680	08.
63	J-wel Wayne	89256	Jne. 20	2— 9—10	10	381.4	12.191	3.19	2.2 –4.2	1.670–1.858	
64	Quoque Molly Deen	89338	Dec. 29	2— 8—18	9	295.2	13.669	4.63	4.1 –5.35	1.846–2.059	
65	Artis Piebe De Kol	90034	July 24	2— 3—11	5	309.2	12.564	4.06	2.75–6.75	1.585–2.283	
66	Aaltje Salo Colantha Mercedes	91268	Jan. 29	2— 3—17	17	355.0	13.243	3.73	3.05–4.55	1.776–1.951	
67	Queen Sarcastic	91687	Jan. 18	2— 4—26	9	332.5	12.127	3.65	2.73–4.4	1.565–1.899	
68	Starie De Pauline Sarcastic	91691	Mar. 17	2—11—19	12	310.2	12.771	3.75	3.4 –4.2	1.670–1.925	
69	Johanna Pauline Paul	95528	May 1	2— 2— 4	15	375.5	15.29	4.19	3.35–4.9	2.015–2.499	
70	Ollie Watson Prima Donna 2d	97564	Mar. 1	2— 1—19	11	400.6	15.672	3.91	3.53–4.43	2.182–2.313	
71	Queen Juliana Dirkje	97608	Apr. 8	2— 2— 8	48	328.9	12.673	3.85	3.2 –4.85	1.655–2.125	
72	Johanna Flora Summers	98479	Jan. 14	2— 1—28	17	328.2	12.239	3.73	2.5 –5.05	1.662–1.752	
73	Salma 2d Pietertje De Kol 2d	98487	Jan. 6	2— 2— 6	11	327.7	12.30	3.95	3.4 –4.65	1.797–1.919	
74	Schroeder Johanna	98779	Dec. 30	2— 1—22	8	279.6	12.675	4.53	3.8 –5.0	1.753–1.968	
75	Netherland Johanna Rue 7th	98780	Jan. 2	2— 1—17	9	343.9	13.571	3.95	3.2 –4.8	1.782–1.994	
76	Johanna Star Piebe	99196	Jan. 12	2— 0—15	23	418.4	13.340	3.19	2.6 –3.8	1.813–1.986	
77	Gewina Piebe Oak	100563	Mar. 3	2— 1—10	6	327.7	12.065	3.66	3.23–4.2	1.664–1.803	
78	Madison Jetske Ormsby	100696	Feb. 28	2—11— 4	7	348.0	13.436	3.86	3.5 –4.6	1.866–2.010	
79	Jessie Fobes tith Violet 4th 2d	100735	Apr. 19	2— 0— 8	11	380.7	15.062	3.96	2.05–4.4	2.005–2.258	
80	Frisby Homestead De Kol	100736	Jne. 7	2— 0—25	9	401.6	13.722	3.41	2.7 –3.95	1.797–2.060	
81	Jessie Fobes Maud Burke 4th	100739	Apr. 25	2— 1—19	9	335.0	13.365	3.99	3.68–4.45	1.798–1.976	
82	Jessie Fobes Sunnyside De Kol	100740	Apr. 28	2— 2— 2	18	382.8	12.385	3.29	2.9 –3.85	1.741–1.886	
83	Jessie Fobes Pessie Homestead	100742	Feb. 6	2— 0— 3	11	408.2	16.012	3.92	3.3 –4.85	2.203–2.381	
84	Daisy Queen De Kol 2d	109940	Jan. 7	2— 2—21	29	383.3	12.480	3.26	2.75–3.7	1.652–1.909	08.
85	Johanna Colantha Daisy	1 4268	Mar. 14	2— 9— 0	11	380.9	16.052	4.21	3.3 –7.98	2.04 –2.457	
86	Madison Julia Pauline	115838	Mar. 11	2— 9— 6	12	412.5	12.488	3.03	2.1 –4.0	1.663–1.864	
87	Madison Haydee	115866	Mar. 11	2—11—21	10	400.3	14.070	3.52	2.85–4.15	1.922–2.066	
88	Piebe Ormsby Ann	116216	Mar. 11	2—10—21	14	350.9	13.613	3.88	3.45–4.8	1.800–2.072	
89	Madrigal Johanna Carmen	116993	Apr. 9	2— 1— 4	11	345.9	15.348	4.44	4.1 –4.9	2.171–2.231	
	Average, 36 tests (36 different cows)			2— 5—27	15	**370.2**	**13.802**	**3.73**	2.1 –7.98	1.565–2.779	
	Average, 110 tests (110 different cows)			2— 4—25	18	**321.3**	**11.149**	**3.47**	1.4 –7.98	814 –2.779	

A Decade of Official Tests of Dairy Cows. 11

Class V—Cows under two years old.

90.	Rogersville Belle Beauty	94572	Jan. 1	1—11—17	64	335.4	11.327	3.38	2.4 -4.75	1.501—1.680
91.	Johanna Clothilde 3d's Canary	97565	A Dr. 2	1—11—15	48	281.8	10.448	3.71	3.05-4.8	1.326—1.755
92.	Uneeda Inka Corndyke B	97769	Dec. 22	1—11—16	9	318.1	10.707	3.37	2.8 -3.95	1.416—1.580
93.	Wisconsin Star	991 3	Jan. 26	1—11—24	16	431.9	14.948	3.46	2.9 -4.15	2.000—2.216
94.	Lilith Locust	100077	Apr. 13	1—10—14	28	308.1	11.733	3.81	2.95-45	1.589—1.773
95.	Wild Rose Homestead	100734	Apr. 4	1—10—28	17	334.5	11.072	3.31	2.9 -3.85	1.522—1.655
96.	Pinehurst Rigtje	101052	Feb. 19	1—11—18	28	374.0	11.670	3.12	2.9 -3.4	1.617—1.741
97.	Lady Longfield 4th's Ormsby 2d	101131	May 6	1—11—23	12	381.5	13.984	3.67	3.25-4.1	1.923—2.051
98.	Grise da Inka	101742	Feb. 16	1—10—29	36	321.3	10.674	3.32	2.6 -4.28	1.373—1.681
99.	Allie Kimball 2d	104432	Apr. 18	1—11— 3	41	360.2	10.686	2.97	2.0 -4.3	1.309—1.749
100.	Anne Battels Beauty 2d	105007	Apr. 17	1—10—24	13	360.8	12.210	3.48	2.7 -4.85	1.638—1.837
101.	De Dikkert Clothilde Pauline	105985	Nov. 9	1—10— 2	24	292.5	10.088	3.45	2.9 -5.25	1.308—1.605
102.	Margery Joice Mutual	106117	Apr. 18	1—11—15	27	301.2	11.091	3.08	3.2 -4.05	1.492—1.662
	Average, 13 tests (13 different cows)			1—11— 6	28	337.8	11.588	3.43	2.0 -5.25	1.308—2.216
	Average, 24 tests (24 different cows)			1—11— 5	29	294.7	10.044	3.43	2.0 -5.25	.645—2.216

Figure 1. Uneeda Netherland Korndyke, 86494 Holstein. Owned by F. J. Bristol, Oakfield, Wis. Production at 2 years 3 months old during 7 days, 18.826 pounds butter fat.

Figure 2. Netherland Johanna Rue 2nd, 58125, Holstein. Owned by C. A. Schroeder, West Bend, Wis. Production at 7 years 11 months old during 7 days, 19.964 pounds butter fat.

Figure 3. Johanna Star Piebe, 99196, Holstein. Owned by E. U. Schultz, Hartford, Wis. Production, at 2 years old, 13.34 pounds butter fat during seven days.

Figure 4. Wisconsin Star, 99198, Holstein. Owned by E. M. Schultz, Hartford, Wis. Production at 2 years old during 30 days, 59.237 pounds butter fat.

Table IV furnishes information as to the herd book numbers of the cows tested, date of beginning of test, age at last calving, days in milk at beginning of test, yield of milk and butter fat during the seven days, average per cent of butter fat during the test, with variations for single milkings; also range in production of butter fat per day during the test. In the last column, the year or years when a cow was previously tested by this Station are given, with the maximum production of butter fat for the seven days if the record made during the past year was not the largest ever reached by the cow. The average results of the tests of the cows included under the different classes are given in the table and, for the sake of comparison with earlier years, similar data are also given in each case for all cows of the different classes that were tested during the past year.

It is of interest to note in comparing the results of the official 7-day tests given in the tables with corresponding tests conducted during the preceding year that 23 tests of mature cows averaging 19.935 pounds of butter fat for seven days were made last year, against 13 tests averaging 21.295 pounds during 1907 to 1908. The credits for production of butter fat for the class ranged from 18.207 to 23.642 pounds for the different cows. Four tests of four different cows came above three pounds of butter fat a day, on the average, for the seven consecutive days and 10 tests came above 20 pounds for seven days. While the results were not quite as high as during the preceding year the records made are very creditable, especially when we consider that the total number of mature cows producing above 18 pounds of butter fat was nearly double that of any previous year.

In class II (cows between four and five years old) the average for the past year was 17.403 pounds (13 different tests), against 17.351 pounds (four different tests) during the previous year. Here there was a variation of from 16.039 to 22.118 pounds in the production of butter fat for seven days. In class III (cows between three and four years old) the average of 18 tests came to 15.251 pounds of butter fat the past year, against 15.876 pounds as the average of 12 tests during 1907 to 1908. In the two other classes the averages came as follows: Class IV (cows between two and three years old) 13.802 pounds, as an average of 36 tests of as many cows, and class V (cows under two years old) 11.588 pounds as the average for 13 tests of 13 different cows. During the preceding year only 16 and 5 records for these classes, respectively, came above the standards of production given (page 7) and were reported in the bulletin. Some 103 of all the tests conducted during the past year came above the standards of production given, against 51 during the year 1907 to 1908. Since the total number of official 7-day tests of Holstein cows conducted during the two years was 248 and 352, for 1907 to 1908 and 1908 to 1909, respectively, it follows that a considerably larger percentage of high records was made during the latter year than in 1907 to 1908, viz., 29.3 per cent, against 20.6 per cent of the total number of tests conducted.

The following records of production of butter fat during seven consecutive days are worthy of special mention among those given in the preceding pages:

Figure 5. Jewel Duchess, 64474, Holstein. Owned by A. L. Williams, Fond du Lac, Wis. Production at 6 years 4 months old, during 7 days, 22.100 pounds butter fat.

Oak DeKol 2nd, 66793, at six years five months, 23.640 pounds.
Piebe Longfield Night, 75749, at four years six months, 22.118 pounds.
Uneeda Netherland Korndyke, 86494, at two years 11 months, 18.826 pounds.
Jessie Fobes Bessie Homestead, 100742, at two years old, 16.012 pounds.
Wisconsin Star, 99193, at one year 11 months, 14.948 pounds.

The records made by these cows place them among the foremost in the list of Holstein cows tested by this Station.

TABLE V.—THIRTY-DAY RECORDS OF HOLSTEIN COWS, 1908-1909.

Number.	NAME OF COW.	Herd book number.	Test began.	Age. Y. M. D.	Days in milk.	YIELD IN THIRTY DAYS. Milk. Lbs.	Fat. Lbs.	PER CENT FAT. Average	Range.	FAT PER DAY. Range. Lbs.
	Class I.—Cows over 5 years old.									
1	Netherland Johanna Rue 2nd	5615	Dec. 30	7—11—11	29	2396.2	82.825	3.46	2.50—5.00	2.05—3.28
2	Olfie Watson Prima Donna	7186	Feb. 16	5— 3— 7	5	2073.0	86.563	4.17	2.90—6.18	2.46—3.18
	Class II.—Cows 4 to 5 years old.									
3	Jessie Gem Inka	7362	Apr. 14	4—11—29	17	2185.2	80.251	3.67	2.93—4.85	2.08—3.44
	Class IV.—Cows 2 to 3 years old.									
4	Quoque Molly Deen	96268	Dec. 28	2— 8—18	8	1397.9	55.394	3.96	2.95—5.35	1.68—2.06
5	Olfie Watson Prima Donna 2nd	97554	Feb. 21	2— 1—19	3	1690.5	62.325	3.69	2.98—4.95	1.93—2.31
6	Schroeder Johanna	98579	Dec. 30	2— 1—22	8	1273.8	50.255	3.95	2.60—5.00	1.41—1.87
7	Netherland Johanna Rue 7th	98790	Dec. 29	2— 1—17	5	1534.0	53.861	3.51	2.35—4.98	1.56—1.99
8	Johanna Star Piebe	99196	Jan. 5	2— 0—15	16	1802.0	55.544	3.08	2.25—3.80	1.75—1.99
9	Jessie Fobes Bessie Homestead	100742	Mar. 28	2— 0— 3	11	1806.4	64.690	3.58	2.90—4.85	1.90—2.38
	Class V.—Cows under 2 years old.									
10	Wisconsin Star	99193	Jan. 16	1—11—24	6	1865.3	59.237	3.29	1.90—4.20	1.71—2.21

TESTS CONDUCTED EIGHT MONTHS OR MORE AFTER CALVING. Six tests of six different cows at eight months or more from calving at the time of testing were conducted during the year. The results of these tests are given in another place for the cows that were awarded prizes for their production at this stage of their lactation.

THIRTY-DAY OFFICIAL TESTS. Ten tests of Holstein cows continued for a period of 30 days were conducted during the year. The owners of the cows and the number of cows tested for each breeder were as follows:

S. B. Jones & Son, Watertown, two cows (Nos. 3 and 9)

Aug. Knospe, Juneau, two cows (Nos. 2 and 5)

C. A. Schroeder & Son, West Bend, four cows (Nos. 1, 4, 6 and 7)

Ed. M. Schultz, Hartford, two cows (Nos. 8 and 10)

The 30-day tests were conducted by the following supervisors: Frank R. Johnston tested two cows, Nos. 2 and 5; Robert Lachmund tested cows Nos. 1, 4, 6 and 7; W. H. Markey tested cows Nos. 8 and 10, and Paul Skeflo tested cows Nos. 3, 8, 9 and 10.

The results of the 30-day tests are given in Table V. The records made are very good in several cases, but do not call for further comment in this place.

TABLE VI.—RETESTS OF HOLSTEIN COWS, 1908–1909

Number		No. of days tested.	Date.	Average Production per Day.		
				Milk.	Butter fat.	Fat.
				Lbs.	Lbs.	Per ct.
1	Jewel Duchess, 64474	7	Mar. 1–8	73.5	3.157	4.30
		1	Mar. 7–8	75.0	2.886	3.85
		1	Mar. 8–9	77.8	2.828	3.63
2	Johanna De Colantha, 52253	7	Mar. 26–Apr. 2	83.1	3.149	3.79
		1	Mar. 31–Apr. 1	82.9	2.930	3.53
		1	Apr. 1–2	84.1	2.971	3.53
3	Jessie Fobes Bessie Homestead, 100742	7	Mar. 28–Apr. 4	58.3	2.287	3.92
		2	Apr. 4–5, Apr. 6–7	63.4	2.206	3.48
		1	Apr. 5–6	62.8	2.298	3.50
4	Oak De Kol 2d, 66793	7	May 3–10	78.2	3.377	4.32
		2	May 7–8, May 9–10	81.4	3.323	4.08
		1	May 8–9	84.9	3.214	3.79

RETESTS. Four retests of cows on official tests were ordered by the Superintendent of the Advanced Registry of the H.-F.

Association and were conducted by this Station during the past year, viz.:

Jewel Duchess, 64744, owned by A. L. Williams, Fond du Lac.

Johanna De Colantha, 52253, owned by W. J. Gillett, Rosendale.

Jesse Fobes Bessie Homestead, 100742, and

Oak De Kol 2nd, 66793, both owned by S. B. Jones & Son, Watertown.

The retests were conducted for 24 hours by the following supervisors in the order given: Roy T. Harris, A. C. Oosterhuis, B. R. Ryall, and Wm. Schwichtenberg. The results of the retest in all cases confirmed those of the official tests, as will be

TABLE VII.—LIST OF WISCONSIN COWS RECEIVING PRIZES OFFERED BY THE HOLSTEIN-FRIESIAN ASSOCIATION OF AMERICA, 1908-1909.

Number.	Name of cow.	Owner.	Age.	Production of butter fat.	Rank of prize.
	Thirty-day records.		Y. M.	Lbs.	
1	Jessie Fobes Bessie Homestead, 100742.	S. B. Jones & Son, Watertown.	2— 0	64.690	2d.
2	Ollie Watson Prima Donna 2nd, 97554.	August Knospe, Juneau.	2— 2	61.003	3d.
3	Wisconsin Star, 99193	E. M. Schultz, Hartford.	2— 0	59.237	5th.
	Seven-day records.				
1	Oak De Kol 2nd, 66793	S. B. Jones & Son, Watertown.	6— 5	23.640	8th.
2	Jewel Duchess, 64474	A. L. Williams, Fond du Lac.	6— 4	22.109	18th.
3	Johanna De Colantha 52253	W. J. Gillett, Rosendale.	8—11	22.044	20th.
4	Ollie Watson Prima Donna, 71767	August Knospe, Juneau.	5— 3	21.780	22d.
5	Piebe Longfield Night, 75749	W. J. Gillett, Rosendale.	4— 5	22.118	3d.
6	Uneeda Netherland Korndyke, 86494.	F. J. Bristol, Oakfield.	2— 3	18.826	2d.
7	Lady Oak 2d's Homestead De Kol, 89107.	S. B. Jones & Son, Watertown.	2—11	17.883	3d.
8	Johanna Colantha Daisy, 114268	L. G. Legler, Juda.	2— 9	16.052	13th.
9	Jessie Fobes Bessie Homestead, 100742.	S. B. Jones & Son, Watertown.	2— 0	16.012	8th.
10	Johanna Pauline Paul, 95528	C. A. Schroeder, West Bend.	2— 2	15.729	10th.
11	Ollie Watson Prima Donna 2d, 97554.	August Knospe, Juneau.	2— 2	15.672	11th.
12	Jessie Fobes 6th's Violet 4th's 2d, 100735.	S. B. Jones & Son, Watertown.	2— 0	15.062	*
13	Wisconsin Star, 99193	E. M. Schultz, Hartford.	2— 0	14.948	16th.
14	Madam De Kol Netherland, 88808	A. L. Williams, Fond du Lac.	2— 4	14.086	22d.
15	Lady Longfield 4th's Ormsby 2d, 101131.	S. B. Jones & Son, Watertown.	2— 0	13.984	*

*Debarred.

TABLE VII — Continued.

Number	Name of cow.	Owner.	Age.	Production of butter fat.		Days between tests.	Prize.
				1st test.	2d test.		
	Seven-day records eight months or more after calving.		Y. M.	Lbs.	Lbs.		
1	Johanna De Kol Wit, 61874.	C. A. Schroeder, West Bend.	5— 8	19.339	14.130	221	2d.
2	Netherland Johanna De Kol 2d, 61871.	C. A. Schroeder, West Bend.	5—11	22.847	11.361	228	6th.
3	Johanna Clothilde 4th, 60986.	University of Wis., Madison.	6— 4	16.315	10.941	153	*
4	Netherland Johanna Rue 2d, 58125.	C. A. Schroeder, West Bend.	6— 6	12.178	10.915	180	8th.
5	Bryonia Woodland K. De Kol, 73198.	F. J. Bristol, Oakfield.	3— 7	18.393	9.906	333	1st.
6	Little Goldie Paul De Kol, 91206.	A. L. Williams, Fond du Lac.	2— 6	13.116	8.146	231	4th.
7	Madam De Kol Netherland, 88808.	A. L. Williams, Fond du Lac.	2— 4	14.086	6.491	236	7th.
8	Johanna Theresa De Kol 2d, 88747.	A. L. Williams, Fond du Lac.	2— 5	11.866	6.017	236	8th.

*Debarred.

seen from Table VI which shows the average daily production of the cows on the official tests and on the days immediately preceding or following the retest, as well as the average daily production on the retest (the latter in heavy type).

PRIZES WON BY WISCONSIN HOLSTEIN BREEDERS. Wisconsin breeders of Holstein cattle, as usual, secured a large number of the prizes offered by the Holstein-Friesian Association during the past year, for records made in 7-day or 30-day tests, or for 7-day records at eight months or more after calving. Table VII gives the names of cows and the owners of the cows, which received prizes in the various classes during 1908–1909, with the production of butter fat during the 7-day or 30-day period. In records made eight months after calving, the yields of butter fat during both the first and the second tests are given, and the number of days between the beginning of the two tests. The production of butter fat on the first test was not considered in the award of prizes for records made eight months after calving.

It will be noted that one first, three each of second and third, and 16 other prizes were awarded to breeders in this state, or 11.2 per cent of the total number of prizes offered by the Association during the year. In addition to the number of prizes given, three records were made that would entitle the owners of the cows to a prize but for the rule of the Association under which no herd may win more than three prizes in any one prize-division.

SEMI-OFFICIAL TESTS OF HOLSTEIN COWS. The records of production of the 35 Holstein cows which completed semi-official yearly tests from the time arrangements were made for the conduct of these tests by this Station in April, 1905, up to the fall of 1908, are given in Bulletins 144, 160 and 172. Only 20 tests coming under this heading were conducted last year. They were made of five cows in the University dairy herd, of which two completed yearly records, viz., Johanna Clothilde 4th, 60986, and Aurora Clothilde, 83719. Table VIII gives information as to the time of commencing these records, the age of the cows, the date of calving and when bred, the number of days in milk and dry during the year when the records were made, the total credits for milk and butter fat, and the average per cent of fat in the milk for the cows during the year.

TABLE VIII.—WISCONSIN HOLSTEIN COWS COMPLETING SEMI-OFFICIAL YEARLY RECORDS, 1908-1909

Record commenced.	Age.	Date of calving.	Bred.	Days in milk.	Production of Milk.	Butter fat.	Per ct. fat.
Aurora Clothilde, 83719.					Lbs.	Lbs.	
Feb. 28, '08..	6—3—17	Feb. 25, '08	July 3, '08	365	16,419.6	572.70	3.49
Johanna Clothilde 4th, 60986.							
July 7, '08..	2—6—28	July 3, '08	Apr. 22, '09	365	11,637.5	435.61	3.74
Average for 35 cows, 1905-8...					13,209.1	457.3	3.46
Average for 37 cows, 1905-9...					13,253.4	459.9	3.47

Of the 37 cows whose yearly records of production have now been determined by this Station on semi-official tests, 11 cows produced more than 500 pounds of butter fat during the year and five more than 600 pounds of butter fat, the average production for all cows being 13,253.4 pounds of milk and 459.86 pounds of butter fat, equivalent to 537 pounds of butter, or nearly 1½ pounds of butter for each day in the year. The records made by Wisconsin Holstein cows in these yearly tests are, therefore, equally creditable to the breeders and to the state and should encourage our Holstein breeders to continue this work on a broader scale than heretofore, in order that their favorites may also in the future prove their capacity for continued dairy production, along with the cows of other dairy breeds that are being tested for an entire year by their respective breed associations.

Tests of Guernsey Cows.

All tests of Guernsey cows conducted last year were semi-official and made in co-operation with the American Guernsey Cattle Club. The rules of this club provide that tests of one day's duration shall be conducted each month. The names of the owners of the cows, the number of monthly tests conducted during the year and the total number of cows entered on these tests will be seen in Table IX.

Table IX.—Tests of Guernsey Cows, 1908-1909.

No.	Name of Owner.	Number of cows tested.	Number of separate tests.
1	J. H. Beirne, Oakfield	10	71
2	M. D. Cunningham, Kansasville	2	5
3	Howard Greene, Genesee Depot	13	87
4	H. W. Griswald, West Salem	8	60
5	W. J. Heid, Ft. Atkinson	4	43
6	Helendale Farms, Athens	13	62
7	J. G. Hickcox, Whitefish Bay	2	9
8	Chas. L. Hill, Rosendale	15	81
9	W. D. Hoard, Ft. Atkinson	25	176
10	M. B. Lee, Tomah	8	17
11	R. Tratt, Whitewater	24	147
12	The University of Wisconsin	8	8
13	Aug. H. Vogel, Nashotah	5	32
14	M. L. Welles, Rosendale	16	121
	Total	143	919

The total number of monthly tests conducted during the year were 919, against 900 during the preceding year. One hundred forty-three different cows were tested, against 137 during 1907 to 1908. The monthly tests of the Guernseys in the University dairy herd were conducted by our regular supervisors, their work being verified by the Dairy Department of the University of Illinois in the same way as in the case of tests of Holstein and Jersey cows in our herd. Fifty-one Guernsey cows completed yearly records during the past year, against 38 during 1907-8. Table X gives the names of the Wisconsin cows which last year qualified for admission to the Advanced Register of the A. G. C. C., with the names of the owners, the age of the cows and their production during the year.

Figure 6. Rex Eldou 2d, 18763, Guernsey. Owned by Chas. L. Hill, Rosendale, Wis. Production at 3 years 11 months old, 386.7 pounds butter fat for one year.

Figure 7. Natzie Martin, 22925, Guernsey. Owned by A. H. Hinman, Allenville, Wis. Production at 4 years 1 month old, 352.6 pounds butter fat for one year.

Figure 8. Wowkle Martin, 23031, Guernsey. Owned by A. H. Hinman, Allenville, Wis. Production at 2 years 5 months old, 333.95 pounds buttter fat for one year.

Figure 9. Careno 2d. 9180, Guernsey. Owned by W. D. Hoard, Fort Atkinson, Wis. Production at 11 years 5 months old, 532.99 pounds butter fat for one year.

TABLE X.—WISCONSIN GUERNSEY COWS ADMITTED TO ADVANCED REGISTER, A. G. C. C. 1908–1909.

	Dropped.	Calved.	Record commenced.	Service.
Owner: J. H. Beirne, Oakfield.				
1. Imp. Hotton's Nellie 3d, 22562	Sept. 20, 1904	Apr. 12, 1908	April, 1908	July 18, 1908
2. Imp. New Grove Queen, 23647	May 24, 1899	Apr. 2d, 1908	May, 1908	July 14, 1908
3. Imp. Belerma 3d, 23649	May 31, 1902	Mar. 28, 1908	April, 1908	July 17, 1908
4. Imp. Lady Amy of Les Choffins, 23655	Apr. 30, 1905	Apr. 4, 1908	April, 1908	Aug. 5, 1908 / Nov. 20, 1908
5. Imp. Dumont Rose 3d, 23657	Oct. 9, 1905	May 28, 1908	June, 1908	Nov. 28, 1908
Owner: M. D. Cunningham, Kansasville.				
1. Fesdo's Susie, 17229	Sept. 18, 1903	Jan. 8, 1908	Jan., 1908	Apr. 25, 1908
2. Elvira Standard, 18575	Dec. 30, 1904	Dec. 8, 1907	Dec., 1907	Apr. 22, 1908
Owner: Howard Greene, Genesee Depot.				
1. Bessie K., 16750	Fall of 1901	Oct. 29, '07 / Sept. 10, '08	Nov., 1907	Nov. 27, 1907
2. Aura of Rosendale, 18621	Nov. 5, 1904	Aug. 21, 1907	Nov., 1907	Mar. 30, 1908
Owner: W. J. Held, Ft. Atkinson.				
1. Lenorill of Oak Grove, 23073	Apr. 25, 1906	June 11, 1908	June, 1908	Dec. 30, 1908 / Jan. 1, 1909
Owner: Helendale Farms, Athens.				
1. Industry, 19306	May 13, 1905	Oct. 17, 1907	Nov., 1907	Mar. 30, 1908
2. Yeksa Unis, 19790	Nov. 14, 1905	Feb. 17, 1908	Mar., 1908	Aug. 2, 1908
3. Lily Yeksa, 19791	Nov. 21, 1905	Feb. 15, 1908	Mar., 1908	May 26, 1908
Owner: Chas. L. Hill, Rosendale.				
1. Rex Eldou 2d, 18763	Nov. 18, 1903	Oct. 20, 1907	Nov., 1907	Mar. 31, 1908
2. Twilight's Valentine, 19310	Feb. 14, 1905	Sept. 7, 1907	Sept., 1907	Jan. 6, 1908
3. Nina of Rosendale, 20118	Oct. 7, 1905	Jan. 22, 1908	Feb., 1907	June 24, 1908
4. Dawn of Rosendale, 20119	Oct. 14, 1905	Jan. 12, 1908	Jan., 1908	June 9, 1908 / July 10, 1908 / Nov. 25, 1908 / Jan. 23, 1909
Owner: A. H. Hinman, Allenville.				
1. Natzie Martin, 22925	Sept. 9, 1903	Sept. 28, 1907 / Aug. 3, 1908	Oct., 1907	Oct. 20, 1907
2. Wowkle Martin, 23031	Feb. 24, 1905	July 20, 1907	July, 1907	Jan. 29, 1908
3. Minette Martin, 23475	Apr. 11, —	Sept. 27, 1907	Oct., 1907	Apr. 11, 1908
Owner: W. D. Hoard, Ft. Atkinson.				
1. Careno 2nd, 9180	Mar. 16, 1896	Aug. 17, 1907	Aug., 1907	Feb. 14, 1908
2. Cinilla, 12894	July 20, 1900	Oct. 22, 1907	Nov., 1907	Feb. 15, 1908 / June 23, 1908 / Aug 7, 1908 / Oct. 7, 1908
3. Lady Lavene, 12896	Sept. 23, 1898	Sept. 18, 1907	Sept., 1907	Feb. 10, 1908 / Mar. 5, 1908
4. Contessa of Atkinson, 12947	Aug. 22, 1898	Feb. 9, 1907	May, 1907	Feb. 20, 1907 / May 20, 1907 / Apr. 29, 1907
5. Corellia, 13931	Nov. 8, 1900	Aug. 25, 1907	Sept., 1907	Jan. 12, 1908 / Feb. 2, 1908
6. Marsena of Prospect, 15506	Feb. 3, 1902	Sept. 19, 1907	Sept., 1907	Feb. 27, 1908
7. Dottie Fern, 15507	Mar. 24, 1902	Oct. 2, 1907	Oct., 1907	Nov. 14, 1907
8. Birdie L., 15522	July 22, 1901	Nov. 8, 1907	Nov., 1907	
9. Bernhardt, 16060	Sept. 12, 1902	Apr. 10, 1907	May, 1907	May 9, 1907 / Nov. 28, 1907
10. Mabel of Prospect, 16061	Sept. 20, 1902	Apr. 29, 1908 / Apr. 4, 1909	May, 1908	June 23, 1908 / July 8, 1909
11. Lady Lavene 2d, 16063	Sept. 26, 1902	Sept. 19, 1907	Sept., 1907	Feb. 25, 1908
12. Contes-a's Favorite, 16724	Apr. 5, 1903	Mar. 5, 1907 / Mar. 15, 1908	May, 1907	May 10, 1907 / June 5, 1907
13. Cremo, 17060	Aug. 27, 1903	Sept. 21, 1907	Oct., 1907	Jan. 13, 1908
14. Helen Kellar, 17265	Jan. 10, 1904	Sept. 10, 1907 / June 18, 1908	Sept., 1907	Jan. 21, 1908 / Feb. 13, 1908 / Aug. 21, 1908
15. Almira, 18384	Jan. 17, 1905	Apr. 30, 1907 / Mar. 25, 1908	May, 1907	June 18, 1907 / May 13, 1908
16. Imp. Duveaux Lass I. 22560	Dec. 8, 1904	June 17, 1907	July, 1907	Jan. 12, 1908 / Feb. 7, 1908

TABLE X.—WISCONSIN GUERNSEY COWS ADMITTED TO ADVANCED REGISTER, A. G. C. C., 1908-1909—Continued.

	Dropped.	Calved.	Record commenced.	Service.
Owner: M. B. Lee, Tomah.				
1. Mabel of Oak Ridge, 15842	Aug. 3, 1902	Oct. 3, 1907	Oct., 1907	Apr. 6, 1907
2. Maud of Oak Hill, 20104	July 3, 1904	Mar. 15, 1908 / Jan. 4, 1909*	Mar., 1908	June 16, 1908 / Mar. 11, 1909
3. Glenwood's Princess of Tomah, 20902	Dec. 21, 1905	Jan. 28, 1908	Feb., 1908	June 26, 1908 / Oct. 20, 1908
Owner: Ralph Tratt, Whitewater.				
1. Troy Center Belle, 14566	Sept. 17, 1901	Feb. 24, 1908	April, 1908	Mar. 29, 1908
2. Tristan's Royalette, 16884	Nov. 7, 1898	Feb. 24, 1908	April, 1908	Sept. 3, 1908
3. Derinna, 1777	Sept. 29, 1901	Apr. 24, 1908	April, 1908	Dec. 15, 1908
4. Janetius, 18087	Aug. 17, 1902	Jan. 13, 1908	April, 1908	
5. Mayflower of Eagle, 18119	July 10, 1904	Mar. 1, 1908	April, 1908	Sept. 1, 1908 / Dec. 12, 1908
Owner: University of Wisconsin, Madison.				
1. Hopeful Mollie, 18768	Apr. 2, 1904	Oct. 16, 1907	Nov., 1907	June 6, 1908 / Sept. 17, 1908
2. Handsome Gypsie, 18769	May 1, 1904	Jan. 15, 1908 / Mar. 25, 1909	Jan., 1908	June 22, 1908
3. Model Countess, 18770	May 18, 1904	Nov. 9, 1907	Nov., 1907	May 24, 1908
Owner: August H. Vogel, Nashotah.				
1. Hedwig B., 12574	Dec. 9, 1899	May 23, 1907	June, 1907	Oct. 30, 1907
2. Augusta's Pride, 13350	Aug. 16, 1900	May 1, 1908	May, 1908	Oct. 1, 1908
3. Augusta's Lily, 16912	Sept. 5, 1903	Nov. 16, 1907	Nov., 1907	Apr. 6, 1908
Owner: M. L. Welles, Rosendale.				
1. Imp. Dolly of La Ramee IX, 22555	Mar. 25, 1905	Oct. 11, 1907	Oct., 1907	Jan. 3, 1908 / May 20, 1908 / Aug. 4, 1908

* Aborted.

Figure 10. Lady Lavene, 12806, Guernsey. Owned by W. D. Hoard, Fort Atkinson, Wis. Production at 9 years old, 524.52 pounds butter fat in one year.

Figure 11. Mabel of Oak Ridge, 15842, Guernsey. Owned by M. B. Lee, Tomah, Wis. Production at 5 years 2 months old, 430.63 pounds butter fat for one year.

Figure 12. Maud of Oak Hill, 20104, Guernsey. Owned by M. B. Lee, Tomah, Wis. Production at 3 years 10 months old, 373.73 pounds butter fat for one year.

Figure 13. Tristan's Royalette, Guernsey. Owned by Ralph Tratt, Whitewater, Wis. Production at 9 years 4 months old, 499.51 pounds butter fat for one year.

Figure 14. Augusta's Lily, 16912, Guernsey. Owned by Aug. H. Vogel, Nashotah, Wis. Production at 4 years 2 months old, 365.72 pounds butter fat for one year.

TABLE XI.—SEMI-OFFICIAL YEARLY TESTS OF WISCONSIN GUERNSEY COWS, OCTOBER 1, 1908, TO OCTOBER 1, 1909.

No.	Name.	Herd book No.	Adv. R. No.	Age.	Pounds milk.	Pounds butter fat.	Per cent butter fat.
1	Careno 2nd	9180	722	11— 5	8,069.70	532.99	5.94
2	Lady Lavene	12896	739	9—	11,235.70	524.58	4.67
3	Imp. Hotton's Nelly III	22562	901	3— 7	8,675.60	510.49	5.89
4	Yeksa Unis	19790	858	2— 3	9,320.20	503.61	5.40
5	Tristan's Royalette	16884	863	9— 4	10,353.80	499.51	4.82
6	Imp. Belerma III	23649	851	5—10	7,853.80	452.37	5.76
7	Lily Yeksa	19791	859	2— 3	8,556.50	450.34	5.26
8	Contessa of Atkinson	12947	723	8— 9	9,492.70	449.66	4.74
9	Corella	13031	724	6—10	8,563.00	448.01	5.23
10	Mable of Prospect	16081	906	5— 8	8,933.40	444.70	4.98
11	Derinna	1777	864	6— 7	9,109.00	442.70	4.86
12	Aura of Rosendale	18621	747	2— 9	8,773.40	442.41	5.04
13	Imp. Duveaux Lass I	22560	625	2— 7	8,265.60	440.63	5.33
14	Cinilla	12894	806	7— 3	9,705.20	439.94	4.53
15	Mabel of Oak Ridge	15842	767	5— 2	9,096.30	430.63	4.75
16	Janetius	18087	860	5— 7	8,756.00	428.33	4.89
17	Imp. New Grove Queen	23647	899	8—11	7,938.50	427.87	5.39
18	Lady Lavene 2d	1,083	737	4—11	8,576.00	427.61	4.99
19	Almira	18384	727	2— 4	8,279.60	426.29	5.15
20	Lady Amy of Les Choffins	23655	900	2—11	7,512.20	416.48	5.54
21	Hedwig B	12574	719	7— 5	9,536.40	415.28	4.35
22	Mabel Countess	18770	748	4— 6	8,890.10	410.20	4.61
23	Bernhart	16080	726	4— 8	7,259.40	408.78	5.63
24	Mayflower of Eagle	18119	861	3— 8	8,518.00	406.22	4.77
25	Marsena of Prospect	15506	736	5— 7	8,250.90	400.09	4.85
26	Troy Center Belle	14566	862	6— 6	7,428.10	396.58	5.34
27	Imp. Dumont Rose III	23657	902	2— 8	8,240.40	391.19	4.75
28	Dottie Fern	15507	804	5— 6	8,512.10	389.02	4.57
29	Rex Eldou 2d	18763	753	3—11	7,980.75	386.70	4.85
30	Birdie L	15522	803	6— 4	6,992.70	384.76	5.50
31	Industry	19308	766	2— 5	6,787.80	379.27	5.59
32	Hopeful Mollie	18768	743	4— 7	7,272.40	378.29	5.20
33	Maud of Oak Hill	20104	892	3—10	6,167.80	373.23	6.05
34	Bessie K	16750	746	6—	7,874.70	371.73	4.72
35	Augusta's Lily	16912	759	4— 2	7,534.00	365.72	4.85
36	Augusta's Pride	13350	876	7— 8	6,421.80	363.94	5.67
37	Lenoril of Oak Grove	23073	903	2— 2	7,191.60	360.76	5.02
38	Contessa's Favorite	16724	721	4— 1	6,805.30	354.35	5.21
39	Natzie Martin	22925	703	4— 1	6,447.70	352.60	5.47
40	Fesdo's Susie	17229	805	4— 4	7,507.40	346.22	4.61
41	Elvira Standard	18575	808	2—11	5,692.70	343.33	6.03
42	Helen Keller	17265	738	3— 8	6,286.10	334.29	5.33
43	Wowkle Martin	23031	730	2— 5	7,150.20	333.95	4.67
44	Cremo	17080	735	4— 1	5,766.50	333.58	5.78
45	Glenwood's Princess of Tomah's	20002	816	2— 1	6,155.10	325.16	5.28
46	Nina of Rosendale	20118	829	2— 3	6,299.60	324.66	5.15
47	Dawn of Rosendale	20119	830	3— 3	6,008.45	323.89	5.39
48	Twilight's Valentine	19310	752	2— 7	5,202.40	318.88	6.13
49	Minette Martin	23475	764	2— 6	6,349.60	317.17	5.00
50	Handsome Gypsy	18769	817	3— 9	6,692.10	315.96	4.72
51	Imp. Dolly of La Ramee IX	22555	797	2— 7	7,416.40	306.64	4.13
	Average			4— 8	7,814.6	399.05	5.11

The average production of the 51 cows that were tested last year by this Station and qualified for the Advanced Register of the A. G. C. C. was 7,814.6 pounds of milk and 399.05 pounds of butter fat, average fat content 5.11 per cent. Since half the number of these cows were under four years of age, this average production must be considered very satisfactory indeed. Twenty-four cows made records of 400 pounds of butter fat or over and four, viz., Careno 2nd, Lady Lavene, Imp. Hotton's Nellie, and Yeksa Unis produced over 500 pounds during the year. While none of the records given exceeded those previously made by Guernsey cows of the respective ages the average production of the cows tested during the year as well as the production of a large majority of the cows, is very creditable and justifies the reputation of the owners as breeders of Guernseys of a large capacity for dairy production.

TESTS OF JERSEY COWS.

Twenty-four Jersey cows were placed on semi-official tests during the year, and three Jerseys, all owned by the University of Wisconsin, were tested for seven consecutive days. The former tests were conducted in conjunction with the American Jersey Cattle Club under the rules adopted for admission to the Register of Merit of this club, and the latter under the direction of the University of Illinois. In addition, seven tests of two or three days' duration were made of the Jersey cow Jacoba Irene, 146443 (see below).

One hundred fifty-four tests were conducted of the cows placed upon semi-official yearly tests, as shown in following list.

No.	Name of breeder.	Number of cows.	Number of separate tests.
1	A. O. Auten, Jerseyville, Ill..................................	1	7
2	J. Q. Emery & Son, Edgerton, Wis........................	5	25
3	E. E. Hill, Tomah, Wis..	1	4
4	Muskego Lakes Jersey Herd, Muskego Lakes, Wis........	14	95
5	The University of Wisconsin, Madison, Wis..............	1	8
6	Wyatt & Son, Tomah, Wis....................................	3	22
		25	161

The names and addresses of the owners of the cows and the number of cows tested for each are also given in this table.

The three official tests of Jersey cows in the University dairy herd were conducted by R. E. Harris of Warrens (No. 1) and J. W. Hayden of the University of Illinois (Nos. 2 and 3). The results of these tests will be seen in Table XII.

The occasion for making the tests of Jacoba Irene, 146443, is given in our last year's bulletin on the dairy tests.[2] This

---

[2] Bulletin 172, p. 30.

cow now holds the Jersey record for production of butter fat for one year, being credited with a production of 17,253.2 pounds of milk, 952.966 pounds of butter fat, average per cent 5.53, for the year ending January 24, 1909. The detailed results of the tests conducted by this Station are given in Table XIII, showing the production of milk for each month and for the day, the percentages of fat and total solids (calculated from the percentages of fat and lactometer readings), and the production of these components. The low percentages of solids and fat in the milk and the low production during the first two days of the July test and the last day of the October test are explained by the fact that the cow was slightly off feed these days. The variations in the daily results for the other tests are no greater than are generally found in the production of milk and butter fat from single cows.

TABLE XII.—SEVEN-DAY RECORDS OF JERSEY COWS, 1908—1909.

No.	Name of Cow.	Herd book number.	Test began.	Age.	Days in milk.	Yield in seven days.		Per cent fat.		Fat per day, range.
						Milk.	Fat.	Average	Range.	
				Y. M. D.		Lbs.	Lbs.			Lbs.
1	Flossy Ella.........	207765	Mar. 2	2— 5—12	7	180.6	8.922	4.94	3.85—9.40	1.096—1.805
2	Just in Time.......	177907	Oct. 23	5— 6—20	18	262.7	14.008	5.33	4.10—6.65	1.845—2.169
3	Macella 3rd.........	149721	Oct. 23	8—10— 1	35	249.1	13.170	5.29	4. 0—7. 0	1.705—2.161

TESTS OF RED POLLED AND GRADE COWS.

Ten 2-day tests and two 3-day tests were made during the past summer and fall. The former tests were conducted at the request of the Secretary of the Red Polled Cattle Club of America, a total of five cows owned by J. W. Martin and H. A. Martin of Gotham and C. L. Underwood of Avoca, having been tested twice in competition for prizes for the production of butter fat on yearly tests offered by the Red Polled Cattle Club. These tests will be continued during the present testing year. The 3-day tests were conducted by L. R. Davies, assistant in dairy tests, at the State Fair grounds at the time of the Fair, September 14 to 16, 1909. Only two cows were entered on the "milk test," for which arrangements were made jointly by the State Board of Agriculture and the Wisconsin Red Polled Breeders' Association, viz.: Lottie B, 22560, owned by A. W. Dopke

TABLE XIII.—MONTHLY TESTS OF JACOBA IRENE, A. J. C. C., 146443, CONDUCTED BY THIS STATION, 1908–1909.

DATE	Morning					Noon					Evening					Total Daily Production					
	Milk	Fat		Total solids		Milk	Fat		Total solids		Milk	Fat		Total solids		Milk	Fat		Total solids		
	Lbs.	Per cent.	Lbs.	Per cent.	Lbs.	Lbs.	Per cent.	Lbs.	Per cent.	Lbs.	Lbs.	Per cent.	Lbs.	Per cent.	Lbs.	Lbs.	Per cent.	Lbs.	Per cent.	Lbs.
1908.																				
July 29	(19.8	4.50	.891	13.90	*2.752)	10.1	3.73	.377	13.10	1.323	16.3	6.23	1.015	15.15	2.469	44.1	4.74	2.091	13.91	6.134
July 30	17.7	3.95	.699	13.23	2.342	9.8	3.18	.312	12.68	1.243	17.5	7.15	1.251	16.23	2.840	45.3	5.07	2.297	14.34	6.495
July 31	18.0	4.08	.734	13.40	2.412	12.4	4.95	.614	14.44	1.791	15.8	8.15	1.288	17.11	2.703	46.6	5.67	2.644	14.94	6.961
Aug. 1	18.4	4.03	.742	13.41	2.467															
Aug. 29	19.1	3.65	.697	13.20	2.521	(18.2	8.80	1.602	17.71	3.223)	6.3	3.80	.239	13.13	.827	42.5	5.30	2.253	14.66	6.233
Aug. 30	15.7	2.63	.413	12.25	1.923	17.1	7.70	1.317	16.87	2.885	6.4	3.95	.253	13.21	.845	40.0	5.31	2.125	14.70	5.879
Aug. 31						17.9	8.15	1.459	17.38	3.111										
Sept. 28	14.9	3.30	.492	13.33	1.954	(16.6	7.80	1.295	16.86	2.799)	5.9	4.73	.279	16.86	.835	37.3	5.43	2.025	14.98	5.586
Sept. 29	18.3	3.15	.576	13.03	2.384	16.5	7.60	1.254	16.75	2.797	2.2	4.63	.102	14.26	.315	36.2	5.57	2.017	15.15	5.484
Sept. 30						15.7	8.53	1.339	17.74	2.785										
Oct. 28	(12.5	3.30	.413	12.93	1.617)	13.5	6.63	.895	15.68	2.117	11.7	9.68	1.133	18.43	2.157	36.9	6.61	2.438	15.74	5.807
Oct. 29	11.7	3.50	.410	13.10	1.533	11.8	6.68	.788	16.04	1.892	7.4	4.85	.359	14.22	1.052	29.3	4.81	1.410	14.26	4.177
Oct. 30	10.1	2.60	.263	12.21	1.234															
Nov. 20	(7.9	2.80	.221	12.30	.972)	11.2	5.43	.608	14.83	1.661	8.3	8.83	.733	18.02	1.496	27.6	5.04	1.556	15.03	4.148
Nov. 21	8.1	2.65	.215	12.23	.991	12.5	6.00	.750	15.13	1.891	7.5	7.10	.533	16.25	1.219	29.4	5.24	1.540	14.43	4.242
Nov. 22	9.4	2.73	.257	12.04	1.132															
Dec. 22	(9.6	2.90	.278	12.29	1.179)	13.1	7.88	1.032	16.66	2.182	5.2	4.70	.244	14.16	.736	28.9	5.44	1.573	14.59	4.217
Dec. 23	10.6	2.80	.297	12.25	1.299	11.1	6.53	.725	15.41	1.711	8.4	9.15	.769	18.13	1.523	27.6	6.16	1.701	15.19	4.192
Dec. 24	8.1	2.55	.207	11.82	.958															
1909.																				
Jan. 22	10.3	4.68	.482	13.81	1.422	(7.5	6.40	.480	15.53	1.165)	7.0	7.83	.548	16.82	1.177	25.3	6.17	1.562	15.25	3.860
Jan. 23	10.4	4.80	.499	14.03	1.459	8.0	6.65	.532	15.76	1.261	6.4	6.85	.438	15.87	1.015	25.3	6.05	1.530	15.0	3.821
Jan. 24						8.5	6.98	.593	15.85	1.347										

* Preliminary milking.

of North Milwaukee, and Lucy 2nd, 22727, owned by Geo. In-Eichen, Geneva, Indiana. The production of these cows during the 3-day tests was as follows:

Lottie B, 98.2 pounds of milk, 3.200 pounds of butter fat, average per cent fat, 3.26.

Lucy 2nd, 105.1 pounds of milk, 3.934 pounds of butter fat, average per cent fat, 3.74.

A single grade cow owned by M. L. Welles of Rosendale, was tested twice during the past year. She did not complete her year's record.

Advanced Registry of the Breed Associations.

The following Table XIV gives the requirements for admission to the Advanced Registers established by the various breed associations. Cows credited with the production given of butter fat or of milk and butter fat on authenticated tests conducted under the direction of an agricultural college or experiment station are admitted to the Advanced Register (or Register of Merit) of the respective breed associations.

Table XIV—Requirements for Admission to the Advanced Registers of Breed Associations.

Age.	Guernsey.		Holstein.	Jersey.		Ayrshire.	
	7-day record.	Year record.	7-day record.	7-day record.	Year record.	Year record.	
	Pounds butter fat.					Lbs. milk and butter fat.	
2 years	10.0	250.5	7.2		260 (2½ yrs.)		
3 "	11.66	287.0	8.8	12.0	300	6,000	214.3
4 "	13.32	323.0	10.4		350	6,500	236
5 "	15.0	360.0	12.0		400	7,500	279
						8,500	322
Requirements increase each day by pounds..	.00456	.1	.00439			1.37 / 2.74	.06* / .12†

*Two-year-old form.
†Three-year-old form.

There is no increase in the requirements for any breed after a cow is five years old. The age of a cow is taken at the beginning of the record, in case of the Guernsey, Jersey, and Ayrshire breeds, while in the case of the Holstein breed it is taken at the time of last calving.

Charges for the Conduct of Dairy Tests

The following schedule gives the charges for conducting dairy tests by this Station:

For 1-day tests.... $6.00 For 7-day tests.... $25.00
For 2-day tests.... 8.00 For 30-day tests.... 80.00

For additional days beyond 2, 7, and 30 days, $3 per day

These amounts cover the entire cost of the tests to breeders so far as this Station is concerned and include all necessary expenses of the supervisors of the tests (traveling, hotel, per diem, etc.). Breeders supply the sulfuric acid used on the tests, pay notary fees and express charges on Babcock testers, convey supervisors to and from nearest railway station and provide for their accommodation at the farm during the tests.

The number of cows that may be tested in a herd at a time varies according to the number of milkings per day. The following statement shows the maximum number of cows that may be placed on official 7 or 30-day tests or on monthly tests at the same time, when the cows are milked two, three or four times a day.

	Milkings per day for one or more cows.	Maximum number tested at a time.
Official tests (7 or more days' duration).	2 or 3	6
	4	5
Monthly tests (1 or 2 days' duration)...	2	10
	3	8
	4	6

The organization of breeders' associations which has recently taken place in different parts of the state has made it possible to reduce materially the cost of Advanced Registry testing for members of the associations, from the fact that supervisors can make these tests without a loss of time between and because of the incident saving in traveling and other expenses. These Advanced Registry tests for members of breed associations and for farmers similarly situated will be conducted at the following rates:

For 2-day tests (required by the H.-F. Assn. and the A. J.
 C. C.) $5.00
For 1-day tests (required by the A. G. C. C.)...........$3.00,

provided at least six breeders in an association take up the work so that the supervisors can make the tests in a circuit without a loss of time between the tests. Arrangements for conducting circuit tests within local breeders associations should be made through the secretary of the association with whom the details and appointments relative to the conduct of tests will be arranged by this department.

The charges for tests conducted in connection with the Wisconsin Dairy Cow Competition will be the same as for the circuit tests. This Competition, which has been rendered possible through the liberality of W. W. Marsh of Iowa and a number of prominent Wisconsin breeders, will be based on records of production of butter fat in semi-official yearly tests conducted between November 1, 1909 and November 1, 1911. A prize fund has been subscribed aggregating $2,000 which will be used as cash prizes of $50 to $500 each for individual and herd records made by Wisconsin cows. Further details concerning the Competition and the rules governing the same are given in circular No. 9 of this Station which will be sent upon request to all residents of the state.

II. A DECADE OF COW TESTING.

It will be of interest to summarize the work done in testing pure-bred dairy cows in this state during the past decade in order to arrive at a somewhat definite opinion as to the value and importance of this work to the breeders of dairy cattle and the dairying industry of our state. For this purpose the tables which follow have been prepared, showing the total number of breeders for whom tests have been conducted, also the number of cows tested during the period stated, the data being arranged by counties. (See Table XV.)

This work has been done in 27 different counties of the state, for 109 different breeders of dairy cattle, 2,764 cows having been tested in all; in addition, about 50 cows have been tested for Michigan and Illinois breeders living near the state line; a large number of miscellaneous tests conducted for only brief periods has not been included in this summary.

It would be a mistake to conclude from these figures that the value of these tests to the dairy industry of our state is confined to the direct benefit these one hundred odd breeders derived from them. As suggested in the introduction of this bulletin, the system of official and semi-official testing of dairy cows is of general importance, since to a large extent our dairy

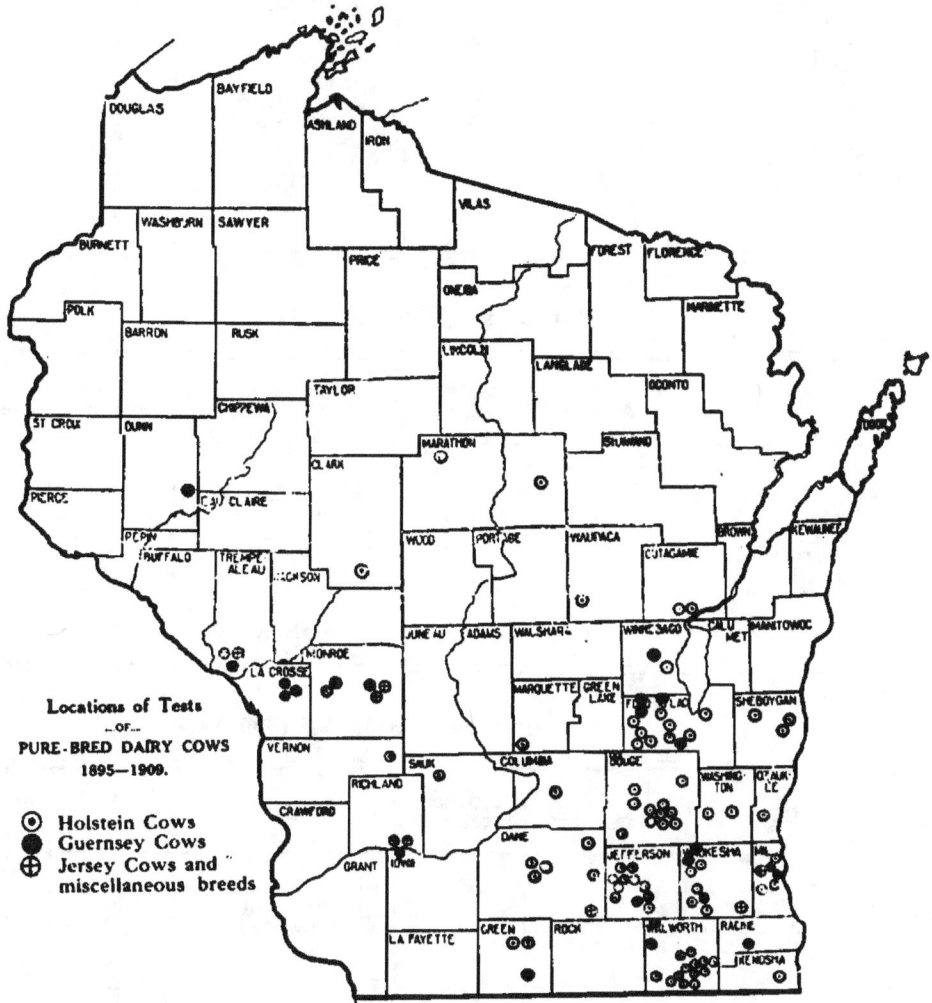

Figure 15. Location of tests of dairy cows of different breeds, which have been conducted in Wisconsin, 1899-1909.

farmers, as well as many in other states, depend on these relatively few breeders to supply them with both males and females. The records made in these tests enable breeders to furnish their customers with authenticated figures for the production of milk and butter fat by the relatives of the stock they sell, or by the cows tested, so that the breeding operations of all are placed on a more solid foundation than would be the case without these tests.

TABLE XV.—SEVEN-DAY HOLSTEIN TESTS, 1899–1909.

County.	No. of farmers.	No. of cows tested.	County.	No. of farmers.	No. of cows tested.
Clark	1	7	Monroe	2	20
Columbia	1	12	Outagamie	2	14
Dane	5	234	Ozaukee	1	2
Dodge	12	571	Sauk	1	2
Fond du Lac	8	261	Sheboygan	3	30
Green	3	102	Vernon	1	3
Jefferson	9	354	Walworth	12	126
Kenosha	1	11	Washington	2	54
Marathon	1	2	Waukesha	6	60
Marquette	1	3	Waupaca	1	22
Milwaukee	2	30	Winnebago	1	10
			Total	76	1,930

GUERNSEY TESTS, 1899–1909.

County.	No. of farmers.	No. of cows tested.	County.	No. of farmers.	No. of cows tested.
Dane	1	29	Monroe	1	9
Dunn	1	22	Racine	1	61
Fond du Lac	3	216	Trempealeau	1	3
Jefferson	3	75	Walworth	2	69
La Crosse	3	19	Waukesha	2	43
Marathon	1	91	Winnebago	1	17
Milwaukee	2	42			
			Total	21	696

TESTS OF OTHER BREEDS, 1899–1909.

County.	No. of farmers.	No. of cows tested.	County.	No. of farmers.	No. of cows tested.
Dane	2	25	Richland	3	11
Fond du Lac	1	62	Trempealeau	2	10
Milwaukee	1	1	Waukesha	1	25
Monroe	2	4			
				12	138

The educational value of the tests should also be kept in mind. The records made by the cows on official tests are, in general, obtained under the most favorable conditions, the cows being liberally fed and well cared for. Hence there is a strong incentive for all to try to obtain similar results in the case of other cows in the herds where the tested cows or their relatives are placed. Wherever these cows go they are, as a rule, the best cows in the herds, and the influence of their blood and of their example tends toward an improvement in the productive capacity of all the cows in the herd and of their offspring. The value of the tests to the dairy industry of this state, cannot, therefore, be measured by the small number of breeders for whom this work has been done so far, or who will directly profit by it in the future, but must be considered from the broader view point of the influence of the tests on the improvement of dairy herds in all parts of the state and in other states.

GROWTH OF DAIRY TESTS. A number of tables are given in the following pages which will show at a glance the growth of the system of testing of dairy cows for the production of milk and butter fat in this state during the past 10 years.[1] The total number of cows tested annually has increased from 132 cows in 1899–00 to 543 in 1908–9, the annual increase during the last three years having been 10, 19 and 30 per cent over the number of tests conducted during the preceding year, or an average of about 20 per cent. The total number of tests conducted has increased during the past decade from 180 to 1,479, the average annual increase during the last three years being 18 per cent.

Figure 16. Relative growth of number of dairy cows tested, 1899-1909.

As these tests ranged in duration from one to 30 days and the number of tests of different lengths vary from year to year, the figures given do not accurately represent the amount of work done in this line during the different years. In order to obtain a correct expression of the growth of this work during the past decade the tests conducted each year have been referred to a 1-day unit, with results shown in Table XVI.

---

[1] The writers acknowledge the assistance of Mr. Paul Skeflo, late supervisor of dairy tests, in the preparation of the diagrams accompanying these tables.

TABLE XVI.—SUMMARY OF DAIRY TESTS, 1899–1909.

	TOTAL NO. OF COWS TESTED.		TOTAL NO. OF SEPARATE TESTS.		TOTAL NO. OF DAYS OF TESTING.		No. of breeders for whom tests were conducted.
		Increase.		Increase.		Increase.	
		Per cent.		Per cent.		Per cent.	
1899–00	132	....	180	....	970	....	17
1900–01	130	— 2	142	— 21	958	— 1	15
1901–02	190	46	373	163	1,565	63	19
1902–03	206	8	451	21	1,892	21	23
1903–04	317	54	748	66	2,805	43	36
1904–05	326	3	864	16	2,682	— 4	42
1905–06	315	— 3	922	7	2,664	— 1	37
1906–07	345	10	969	5	2,740	3	41
1907–08	418	19	1,327	37	3,343	22	46
1908–09	543	30	1,479	11	4,094	22	65

These figures show that the tests conducted during the first year of the period here considered aggregated 970 days of testing and those during the last year 4,094 days. The increase in the total amount of work done expressed in units of 1-day tests during the various years ranged from small negative figures during the second, seventh and eighth year, to 63 per cent in 1901–2 (owing to the commencement of the Guernsey semi-official tests at this time) and 22 per cent during each of the last two years. These figures and others given in the tables in this part of the bulletin are represented graphically in the diagrams.

Figure 17. Increase in total number of tests of dairy cows, 1899–1909.

The number of breeders for whom tests were conducted during the years here considered ranged from 15 in 1900–1 to 65 during the past year. Most of the breeders have had cows in their herds tested during a series of years; in fact, in a few cases during every year since this work was taken up by our Station.

Hence the figures previously given for the total number of breeders for whom tests have been conducted is not greatly in excess of the figure for last year when the largest number of breeders availed themselves of the opportunity to obtain authenticated records of production of their cows which the dairy tests offer.

Figure 18. Increase in number of breeders of dairy cows for whom tests were conducted, 1899-1909.

The number of Guernsey and Holstein tests conducted annually during the past year will be seen from Table XVII. The table shows an increase in the former tests from 42 in 1899-0 to 919 in 1908-09, and in the latter from 127 in 1899-0 to 382 in 1908-9. The figures given in the table need not be discussed in detail. The change in the charges of the American Guernsey Cattle Club for Advanced Registry testing in 1905 is responsible for the decrease in the number of tests conducted during the

TABLE XVII.—SUMMARIES OF GUERNSEY AND HOLSTEIN TESTS, 1899-1909.

YEAR.	No. of Guernsey Tests.			No. of Holstein Tests.		
		Increase.	Decrease.		Increase.	Decrease.
		Per cent.	Per cent.		Per cent.	Per cent.
1899-00	42	....	....	127	....	....
1900-01	10	....	—76	132	4	....
1901-02	107	970	....	192	45	....
1902-03	180	68	....	171	....	—11
1903-04	458	154	....	265	55	....
1904-05	550	20	....	258	....	—3
1905-06	367	....	+33	420	64	....
1906-07	586	60	....	356	....	—15
1907-08	900	54	....	349	....	—2
1908-09	919	2	....	383	10	....

following year, and the large annual increase during the two years 1906–8 probably explains the relatively small gain in the number of Guernsey tests conducted during 1908–9.

Figure 19. Increase in number of tests of Guernsey cows, 1899-1909.

The number of applications for conducting dairy tests received annually from farmers is dependent upon many factors, some of these affect the individual farmer only, others like the character of the season, the quality of the hay crop, or of the corn and silage available for feeding, the prices of feed and of stock, etc., affect all breeders more or less and determine to a large extent the amount of testing we are called upon to do. Some of the minus signs in the table for Holstein tests are very likely traceable to one or more of the general causes suggested.

Figure 20. Increase in number of tests of Holstein cows, 1899-1909.

IMPROVEMENT IN DAIRY PRODUCTION.

In reviewing the results of the work of testing dairy cows done by the Station during the past decade the question naturally arises whether any improvement in the average production of the cows is to be noticed from year to year during this period and if so, whether in the quantity or the quality of the milk secretion, or both, and to what cause or causes the improvement, if there be any, is attributable. The best material for studying this question is found in the results of the 7-day official Holstein tests since these were conducted in a similar manner throughout the period considered and, as it would appear, with a sufficient number of animals to render the averages of value in this study.

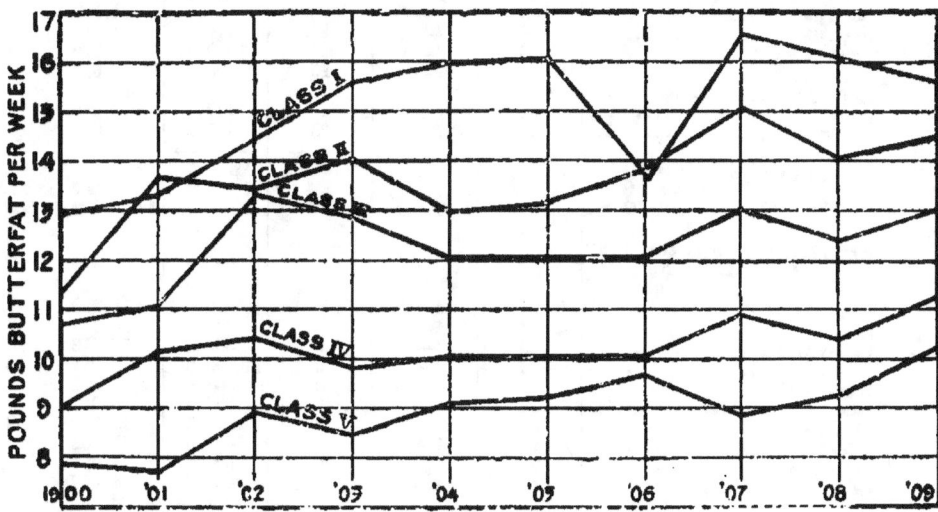

Figure 21. Average production of butter fat by Holstein cows of the various classes on seven-day official tests, 1899-1909.

The average data for the respective classes of Holstein cows tested during each year of the past decade are presented in the following tables, the total number of tests conducted each year being given, the average age of the animals tested, the average distance from calving at the beginning of the tests, and the average production of milk and butter fat for the seven days, with average percentage of butter fat in the milk (i. e. total fat ÷ total milk x 100). In order to trace the tendency of the results toward an improvement, or otherwise, the averages for the first and the second five-year periods have been calculated for each class and are given at the end of the tables. These figures are the arithmetical means of the yields for the various years and the fact that the number of cows tested varied considerably is not taken into consideration, each year being considered a unit and of equal importance. The diagrams show at a glance the general trend of the variations in the production of milk and butter fat.

TABLE XVIII.—AVERAGE 7-DAY RECORDS OF WISCONSIN HOLSTEIN COWS, BY YEARS, 1899-1909.

YEAR.	Number of tests.	Average age.	Average days in milk.	Average production in seven days.		
				Milk.	Butter.	Fat.
Class I—Cows five years old or over.		Y. M.		Lbs.	Lbs.	Per ct.
1899-1900	50	6— 7	33	371.9	12.890	3.47
1900-1901	44	8— 1	33	379.5	13.244	3.49
1901-1902	55	7— 1	20	401.5	14.275	3.56
1902-1903	65	7— 3	26	420.4	14.507	3.45
1903-1904	74	7— 5	22	414.0	14.816	3.58
1904-1905	58	7— 0	19	419.4	14.978	3.57
1905-1906	51	7— 5	25	408.4	13.680	3.35
1906-1907	44	7— 2	27	445.8	16.653	3.75
1907-1908	61	7— 0	26	449.1	16.084	3.58
1908-1909	112	7— 7	23	441.8	15.590	3.53
Average 1899-'04			27	397.5	13.946	3.51
" 1904-'09			24	432.9	15.397	3.56
" 1899-'09			25	415.2	14.672	3.53
Class II—Cows between four and five years of age.						
1899-1900	8	4— 2	56	310.6	11.250	3.62
1900-1901	16	4— 4	24	372.2	13.475	3.62
1901-1902	29	4— 6	38	380.6	13.412	3.52
1902-1903	23	4— 6	22	400.0	13.941	3.49
1903-1904	41	4— 5	48	362.4	12.677	3.50
1904-1905	23	4— 5	21	387.5	13.041	3.37
1905-1906	25	4— 6	29	383.6	13.739	3.58
1906-1907	25	4— 5	24	430.5	15.202	3.53
1907-1908	28	4— 4	24	399.6	14.054	3.52
1908-1909	44	4— 5	20	410.9	14.466	3.52
Average 1899-'04			38	365.2	12.951	3.55
" 1904-'09			24	402.4	14.100	3.50
" 1899-'09			31	383.8	13.526	3.52
Class III—Cows between three and four years of age.						
1899-1900	29	3— 5	32	313.8	10.560	3.38
1900-1901	24	3— 4	31	317.8	10.770	3.39
1901-1902	44	3— 6	23	370.2	13.253	3.58
1902-1903	32	3— 6	41	369.7	12.621	3.41
1903-1904	34	3— 4	17	340.6	11.915	3.50
1904-1905	41	3— 7	31	346.8	11.881	3.43
1905-1906	36	3— 6	28	356.0	11.995	3.37
1906-1907	25	3— 4	36	375.7	12.902	3.43
1907-1908	48	3— 5	36	353.9	12.303	3.48
1908-1909	56	3— 5	25	377.9	12.849	3.40
Average 1899-'04			29	342.4	11.824	3.45
" 1904-'09			31	362.1	12.386	3.42
" 1899-'09			30	352.2	12.105	3.44
Class IV—Cows between two and three years of age.						
1899-1900	21	2— 5	42	260.6	8.811	3.38
1900-1901	27	2— 5	40	292.1	10.043	3.44
1901-1902	50	2— 4	55	293.4	10.235	3.49
1902-1903	39	2— 4	22	279.8	9.697	3.46
1903-1904	66	2— 5	44	290.4	9.994	3.45
1904-1905	84	2— 4	29	294.6	10.068	3.42
1905-1906	48	2— 4	29	296.9	9.996	3.37
1906-1907	60	2— 4	30	301.6	10.731	3.56
1907-1908	84	2— 4	28	299.6	10.385	3.47
1908-1909	110	2— 5	18	321.3	11.149	3.47
Average 1899-'04			41	283.3	9.756	3.45
" 1904-'09			27	302.8	10.466	3.46
" 1899-'09			34	293.0	10.111	3.45

TABLE XVIII.—AVERAGE 7-DAY RECORDS OF WISCONSIN HOLSTEIN COWS, BY YEARS, 1899-1909—Continued.

YEAR.	Number of tests.	Average age.	Average days in milk.	AVERAGE PRODUCTION IN SEVEN DAYS.		
				Milk.	Butter.	Fat.
Class V—Cows below two years of age.		Y. M.		Lbs.	Lbs.	Per ct.
1899-1900	13	1—10	114	243.8	7.843	3.22
1900-1901	9	1— 9	26	241.3	7.770	3.22
1901-1902	14	1—10	18	242.1	8.480	3.50
1902-1903	12	1—11	39	252.5	8.403	3.33
1903-1904	19	1—10	21	263.8	8.990	3.41
1904-1905	20	1—11	35	279.0	9.204	3.30
1905-1906	16	1—11	25	288.1	9.082	3.36
1906-1907	19	1—11	33	279.5	8.934	3.20
1907-1908	19	1—11	34	278.5	9.216	3.31
1908-1909	24	1—11	29	294.7	10.094	3.43
Average 1899-'04			44	248.7	8.297	3.34
" 1904-'09			31	284.9	9.426	3.31
" 1899-'09			37	266.3	8.862	3.33

From Table XVIII it appears that there is an unmistakable upward tendency of the results obtained on these tests from year to year, showing that the average production of the cows tested has increased both as regards the yield of milk and of butter fat during the period considered, *e. g.* in the aged class the average production of the cows during the former 5-year period was 397.5 pounds of milk and 13.946 pounds of butter fat (average per cent, 3.51) and in the latter 432.9 pounds of milk and 15.397 pounds of butter fat (average per cent 3.56), an increase of 9 and 10 per cent, respectively. In the same way class V, heifers below two years of age, produced on the average 248.7 pounds of milk and 8.297 pounds of butter fat (average per cent 3.34) during 1899-04 and 284.9 pounds of milk and 9.426 pounds of butter fat (average per cent 3.31) during the five years, 1904-09, an increase of 15 and 14 per cent, respectively. The average data for all classes were as follows:

1899-04, 327.4 pounds of milk, 11.357 pounds butter fat, average per cent 3.47.

1904-09, 357.0 pounds of milk, 12.355 pounds butter fat, average per cent 3.46.

These figures show an increase in the production of both milk and butter fat of about nine per cent during the second 5-year period over that of the first 5-year period. We note in passing that the average production of all the Holstein cows on the 7-

day official tests included in these tables (1,999 different tests) was 342.2 pounds of milk and 11.855 pounds of butter fat (average per cent 3.46), or an average daily yield of 48.9 pounds of milk and 1.694 pounds of butter fat (equivalent to about two pounds of commercial butter per head daily).

The improvement noted would appear still more marked if the actual averages for all tests conducted during the two 5-year periods were considered instead of the arithmetical means, since more 7-day tests were conducted during the latter 5-year period than during 1899 to 1904, viz., 1,161 against 838. It is evident, therefore, that the conclusion drawn is fully justified by the evidence at hand.

TABLE XIX.—SUMMARY OF RECORDS OF HOLSTEIN TESTS, 1899-1909.

Year.	Holsteins producing over 16 lbs. fat.	Per cent of total number tested.	Holsteins below 3 years making over 12 lbs. fat.	Per cent of total number tested.
1899-00	5	4	1	3
1900-01	8	7	7	19
1901-02	22	11	8	14
1902-03	25	15	6	13
1903-04	25	11	10	12
1904-05	24	10	10	10
1905-'06	13	7	13	22
1906-07	44	23	12	16
1907-08	32	13	15	16
1908-09	62	18	39	29
Average 1899-'04		**10**		**13**
Average 1905-'09		**14**		**15**

A consideration of the proportion of exceptionally high records in the various classes during the two halves of the past decade will lead to a similar deduction as just stated. We will take, for instance, the data for records of production of 16 pounds of butter fat among the 7-day tests conducted during the various years. In 1899 to 1904 five records out of a total of 127 exceeded this amount, or 4 per cent of the total number; in 1900 to 1901 eight out of 120, 7 per cent, in 1901 to 1902, 22 out of 192, 11 per cent, etc. See table XIX; also diagrams, figures 22 and 23. The average per cent for the years 1899 to 1904 came to 10 per cent and for the years 1904 to 1909 to 14 per cent.

In the same way, if the data for the tests of heifers conducted during 1899 to 1909 be considered we have the results given in table XIX, figure 23, showing that 13 per cent of the tests of heifers (classes 4 and 5) come above a production of 12 pounds of butter fat for seven consecutive days during 1899 to 1904, against 15 per cent during 1904 to 1909. While the increase in the production of the heifers tested during the past decade is not very marked, it is too large to be accidental and considered in connection with the results previously referred to, shows that without a doubt the past decade has witnessed a gradual improvement in the productive capacity of cows placed on official tests in this state.

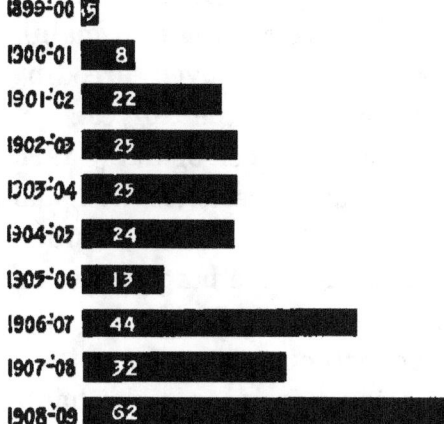

Figure 22. Increase in number of records made by Holstein cows of 16 pounds butter fat or more on seven-day official tests. 1899-1909.

CAUSES FOR IMPROVEMENT. The question next arises as to the cause or causes of the improvement noted. Is it due to improved breeding or to environmental influences, *i. e.*, to the fact that the farmers have learned better to feed and care for their

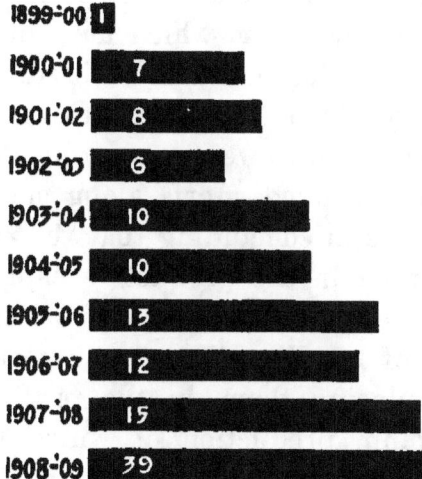

Figure 23. Increase in number of records made by Holstein cows less than three years of age of 12 pounds butter fat or more on seven-day official tests. 1899-1909.

cows and know more now about how to handle the cows so as to reach maximum results than at the beginning of the period considered? On account of the many factors that have a bearing on the records of production obtained in official dairy tests and the relatively brief period covered by the tests here summarized, no definite conclusion can be drawn as to the cause or causes of the improvement in production noted; a general survey of the various factors influencing the production of the cows indicates, however, that the improvement is due mainly to better management and may be due to improved breeding as well. Up to the time of the beginning of the 10-year period considered in this bulletin only a small number of dairy farmers and breeders in this state had had any experience with official testing of cows for milk and butter fat production, and little was known among most of them as to how the best results might be reached in placing cows on these tests. Gradually a more complete understanding of the important points as to the care and feeding of the cows and especially as regards the handling of the cows preliminary to freshening, was gained by more and more breeders, as the years went by, and the results are doubtless to be seen in the higher records made during the latter half of the 10-year period than at its beginning. On the other hand, since it is a well-established law of heredity that "like begets like," we are justified in also seeing in the improvement in production, an evidence of the influence of improved breeding; the cows tested during the last years of the period were, in general, bred from stock that had shown a larger capacity for dairy production than that of the cows from which most of the animals tested in the earlier part of this period were bred. As is the case with variations in the fixed characteristics of any class of animals, dairy improvement is, however, brought about only slowly, as a result of long-continued efforts along specific lines and we feel justified therefore in concluding that the main cause of the improvement observed in the productive capacity of the cows tested came as the result of the adoption of more careful and intelligent methods of handling the cows, especially as regards their feeding and placing them on the test in a prime condition for the maximum production of which they were capable. The same progressive influences will be at work in the future as in the past, but with the high standard of production now reached (see page 44), it cannot be expected that a similar improvement will take place as during the past decade. There will doubtless be reverses during single years and apparently little or no progress at times. There is, however, no reason to doubt that when another 10-year period shall have passed by, the efforts of our breeders will be similarly rewarded as in the past and that the average results obtained in the dairy tests will be still higher than those that were reached during the past few years.

Some Records of Production of Wisconsin Dairy Cows, 1899 to 1909.

A summary of records made by pure-bred cows tested by this Station since this work became of much importance will be of interest to many dairy farmers and breeders. The following table gives the maximum records of production of cows of the three main dairy breeds tested by this Station on semi-official or official tests during the past decade.

The data presented in the following pages show that during the period which has passed since systematic testing of purebred cows was commenced in this state, 11 mature Guernsey cows have produced over 600 pounds of butter fat on semi-official yearly tests and 12 Guernsey cows below four years of age produced over 500 pounds of butter fat. Two Jersey cows and 11 Holsteins (all tested during the past three years), furthermore, produced over 500 pounds of butter fat on semi-official yearly tests.

Considering the results of official tests given in Table XX we note that 17 Holstein cows, four years old or over, produced over 21 pounds of butter fat in seven consecutive days and 90 produced over 18 pounds of butter fat during this period. Of Holstein cows under four years old, 23 produced more than 16 pounds of butter fat, and 119 produced more than 14 pounds during the seven consecutive days. The following cows stand first in their respective classes among the cows tested by our Station:

The Guernsey cow *Yeksa Sunbeam*, 15439, produced during one year, at nine years six months old, 857.15 pounds of butter fat.

The Guernsey heifer *Yeksarose*, 16610, produced during one year, at two years seven months, 638.49 pounds of butter fat.

The Jersey cow *Double Time*, 157531, produced during one year, at nine years two months old, 691.30 pounds of butter fat.

The Holstein cow, *Colantha 4th's Johanna*, 48577, produced

during one year, at eight years two months old, 998.26 pounds of butter fat; during seven consecutive days 28.176 pounds of butter fat, and during 30 consecutive days 110.83 pounds of butter fat.

The Holstein cow *Netherland Johanna De Kol 3d*, 73311, produced on a 7-day official test, at three years 11 months, 19.806 pounds of butter fat.

Nearly all of these cows, with the exception of the last one given, until recently stood first in the respective classes of American cows of the various breeds, as shown by the records published by the different breed associations; at the present writing, Yeksarose's 2-year old record and Colantha 4th's Johanna's yearly and 30-day records still stand as the largest production of butter fat made by a cow of their respective breeds during the periods given.

TABLE XX.—RECORDS OF PRODUCTION OF WISCONSIN COWS, 1899–1909, SEMI-OFFICIAL YEARLY TESTS.

No.	Name.	Age.	Butter fat.
	I.—*Guernsey cows over 4 years old; production over 600 pounds fat.*	Y. M.	Pounds.
1	Yeksa Sunbeam, 15439	9— 6	857.15
2	Lily Ella, 7240	5— 0	782.16
3	Standard's Morning Glory, 12801	4—11	714.01
4	Lilyita, 7241	5— 0	710.53
5	Queen Deette, 9794.[1]	8— 3	669.82
6	Yeksarose, 16610	4— 2	678.16
7	Yeksa Lind, 14275	4— 2	650.56
8	Lily Berkshire, 8110	9—11	631.65
9	Electricia, 9786	8— 1	613.39
10	Gypsy of Racine, 9639	9— 1	611.40
11	Peach O. K., 12994	6— 9	603.64
	Guernsey cows under 4 years old; production over 500 pounds fat.		
1	Yeksarose, 16610	2— 7	638.49
2	Robline 2d, 16117	3— 9	603.59
3	Lily of Helendale, 16915	2— 9	600.49
4	Countess Fantine, 14730	3—11	582.33
5	Riggolette, 16611	2— 7	548.25
6	Penthesella, 17625	2— 3	539.07
7	Yeksalulu Lady, 15583	2— 4	529.69
8	Matinee, 16916	2— 6	524.98
9	Imp. Hotton's Nelly III, 22562	3— 7	510.49
10	Danusia, 16917	2— 5	508.90
11	Yeksa Unis, 19790	2— 3	503.61
12	Primrose of Salem, 12524	3—10	500.63
	II.—*Jersey cows; production over 500 pounds of fat.*		
1	Double Time, 157531	9— 2	691.30
2	Loretta D., 141708.[2]	5— 5	518.70

[1] 16.218 pounds fat in seven days at four years old. Yearly test conducted by Massachusetts Experiment Station.
[2] 16.648 pounds fat in seven days at five years old, and 280.161 pounds in 120 days at seven years eight months old. (St. Louis, 1904.)

No.	Name.	Age.		Butter fat.
	III—Holstein cows; production over 500 pounds fat.	Y.	M.	Lbs.
1	Colantha 4th's Johanna, 48577	8	2	998.36
2	Johanna Bonheur, 60987	5	5	714.25
3	Lady Ormsby, 64352	5	2	651.39
4	Johanna Colantha, 48578	8	5	638.03
5	Johanna DeKol Wit, 61874	4	1	626.15
6	Netherland Johanna DeKol 2d, 61871	4	5	590.46
7	Queen Ormsby, 65901	3	3	584.08
8	Johanna Clothilde 4th, 60986	6	4	572.70
9	Melva Parthena, 50025	7	8	542.07
10	Netherland Bessie, 359	13	3	518.51
11	Johanna Rue DeKol, 58549	4	7	516.21

OFFICAL TESTS OF HOLSTEIN COWS.

Cows over 4 years old; production over 21 pounds fat in 7 days.

No.	Name.	Age.		Butter fat, pounds	
		Y.	M.	7 days.	30 days.
1	Colantha 4th's Johanna, 48577	8	2	28.176	110.83
2	Johanna Colantha 2d, 60891	5	11	26.309	107.74
3	Oak DeKol, 51297	9	2	24.037	
4	Oak DeKol 2d, 66793	6	5	23.640	
5	Netherland Johanna DeKol 2d 61871	5	11	22.847	
6	Queen Ormsby, 65901	5	1	22.691	91.52
7	Jessie Fobes 6th's Homestead, 64296	5	11	22.596	92.20
8	Lady Ormsby, 64352	5	2	22.262	87.38
9	Plebe Longfield Night, 75749	4	6	22.118	
10	Jewel Duchess, 64474	6	4	22.109	
11	Johanna De Colantha, 52263	8	11	22.044	
12	Ollie Watson Prima Donna, 71767	5	3	21.780	86.563
13	Diomandia Dio, 57058	6	9	21.740	
14	Ida Lotta, 50027	8	11	21.689	
15	Winnie Wartena Netherland, 37630	10	7	21.419	88.71
16	Johanna Colantha, 48578	8	5	21.185	
17	Grace Fayne 2d, 44124	7	2	21.037	85.86

90 cows produced over 18 pounds of butter fat in seven days.

No.	Name.	Age.		Butter fat, Pounds	
	Cows under 4 years old; production over 16 pounds butter fat in 7 days.	Y.	M.	7 days.	30 days.
1	Netherland Johanna De Kol 3d, 73311	3	—11	19.806	
2	Uneeda Netherland Korndyke, 86494	2	—11	18.826	
3	De Kol Douglas, 50677	3	—11	18.732	74.00
4	Bryonia Woodland Korndyke De Kol, 73198	3	— 7	18.393	
5	Bessie De Kol Rue 2d, 79374	2	— 5	18.082	70.90
6	Lady Oak 2d's Homestead De Kol, 89107	2	—11	17.883	
7	Gracia Ward, 49517	3	— 9	17.655	
8	Canary Longfield, 59190	3	—11	17.350	72.52
9	Alkartra Polkadot, 50798	3	— 1	17.280	69.35
10	Liscomb Aaggie 3d, 47553	3	—11	17.267	
11	Netherland Johanna Rue 6th, 79373	2	— 5	17.226	66.65
12	De Kol Douglas 5th	3	— 6	17.200	
13	Aaltje Salo Netherland Mechthilde, 78488	3	— 2	17.029	
14	Bessie Ward De Kol, 86553	3	— 2	16.967	
15	Colantha 4th's Sarcastic, 71014	2	—11	16.901	66.44
16	Abbie Douglas De Kol 2d, 65690	3	—11	16.678	68.13
17	Jessie Fobes 6th's Violet 4th, 83263	3	— 0	16.519	67.82
18	Skylark De Kol Johanna, 62231	2	—11	16.514	
19	Lady Longfield 4th's Homestead De Kol, 89109	3	— 0	16.257	
20	Netherland Johanna De Kol 2d, 71871	3	— 4	16.117	
21	Madrigal Josephine Gerben, 82908	3	— 3	16.057	
22	Johanna Colantha Daisy, 114268	2	— 9	16.052	
23	Jessie Fobes Bessie Homestead, 100742	2	— 0	16.012	64.690

119 cows below 4 years old produced over 14 pounds of fat in 7 days.

THE UNIVERSITY OF WISCONSIN

Agricultural Experiment Station

STATION STAFF

The President of the University
H. L. Russell, Director

S. M. Babcock, Assistant Director
Ida Herfurth, Executive Clerk

W. A. Henry, Emeritus Professor of Agriculture

A. S. Alexander, Veterinary Science, In charge of Stallion Licensing
S. M. Babcock, In charge of Agricultural Chemistry
L. J. Cole, In charge of Experimental Breeding.
E. J. Delwiche, Supt. Northern Sub-Stations, (Ashland, Wis.)
E. H. Farrington, In charge of Dairy Husbandry
J. G. Fuller, Animal Husbandry
J. G. Halpin, In charge of Poultry Husbandry
E. B. Hart, Agricultural Chemistry
E. G. Hastings, Agricultural Bacteriology
K. L. Hatch, Agricultural Education; Secretary Agricultural Extension
G. C. Humphrey, In charge of Animal Husbandry
L. R. Jones, In charge of Plant Pathology
E. R. Jones, Soils
C. E. Lee, Dairying

Abby L. Marlatt, In charge of Home Economics
E. V. McCollum, Agricultural Chemistry
J. G. Moore, In charge of Horticulture (Pro tem.)
R. A. Moore, In charge of Agronomy
C. P. Norgord, Agronomy
C. A. Ocock, In charge of Agricultural Engineering
D. H. Otis, Farm Management
M. P. Ravenel, In charge of Agricultural Bacteriology
J. L. Sammis, Dairy Husbandry
J. G. Sanders, In charge of Economic Entomology.
John Spencer, Veterinary Science.
C. W. Stoddart, Soils
H. C. Taylor, In charge of Agricultural Economics
A. R. Whitson, In charge of Soils
F. W. Woll, In charge of Feed and Fertilizer Inspection; Dairy Tests

G. H. Benkendorf, Dairy Husbandry
Emily M. Bresee, Feed and Fertilizer Inspection
L. R. Davies, Dairy Tests; Feed and Fertilizer Inspection
E. F. Eldredge, Agricultural Bacteriology
C. S. Hean, Agricultural Library
Leona Hope, Home Economics
J. Johnson, Horticulture
J. C. Jurrjens, Feed and Fertilizer Inspection
F. Kleinheinz, Animal Husbandry
Alice Loomis, Home Economics
O. G. Malde, Cranberry Investigations, (Grand Rapids, Wis.)
J. C. Marquis, Agricultural Editor
J. G. Milward, Horticulture
W. E. Morris, Feed and Fertilizer Inspection

J. M. Napier, Agronomy
C. R. Orton, Plant Pathology.
P. P. Peterson, Soils
W. H. Peterson, Agricultural Chemistry
A. J. Rogers, Jr., Horticulture; In charge of Nursery Inspection
F. J. Sievers, Soils
H. Steenbock, Agricultural Chemistry
W. W. Sylvester, Agricultural Engineering
A. L. Stone, Agronomy; In charge of Seed Inspection
J. L. Tormey, Animal Husbandry
W. E. Tottingham, Agricultural Chemistry
E. Truog, Soils
H. L. Walster, Soils
W. W. Weir, Soils
F. White, Agricultural Engineering
W. H. Wright, Agricultural Bacteriology

FARMERS' INSTITUTES

George McKerrow, Superintendent
Nellie E. Griffiths, Clerk

The bulletins of this Station are sent free to residents of the State. Names will be entered on the regular mailing list upon request.

www.ingramcontent.com/pod-product-compliance
Lightning Source LLC
Chambersburg PA
CBHW062340220526
45469CB00008B/2787